The New Communications Technologies

Second Edition

The New Communications Technologies

Second Edition

Michael M. A. Mirabito

With contributions by
Barbara L. Morgenstern

Focal Press

Boston   London

Focal Press is an imprint of Butterworth–Heinemann.

Copyright ©1994 by Butterworth–Heinemann
℞ A member of the Reed Elsevier group

Some products are claimed as trademarks by their manufacturers.
Where a trademark appears in this work, and Butterworth–
Heinemann was aware of the trademark claim, the product
has been printed with an initial capital letter.

Recognizing the importance of preserving what has been written, it is
the policy of Butterworth–Heinemann to have the books it publishes
printed on acid-free paper, and we exert our best effort to that end.

Library of Congress Cataloging-in-Publication Data

Mirabito, Michael M., 1956–
    The new communications technologies/Michael M.A. Mirabito with
contributions by Barbara L. Morgenstern. —2nd ed.
        p. cm.
    Includes bibliographical references and index.
    ISBN (invalid) 240-80180-6
    1. Telecommunication systems—Technological innovations.
I. Morgenstern, Barbara L. II. Title.
TK5101.M545 1994
384—dc20   93-35387
                                                                CIP

Butterworth–Heinemann
313 Washington Street
Newton, MA 02158

10 9 8 7 6 5 4 3 2

Printed in the United States of America

To our families
To the men and women of the space shuttle Challenger
and all their fellow travelers.
May their dream live forever in the minds, hearts, and works
Of the present and all future generations.

CONTENTS

Preface

INTRODUCTION

Since *The New Communications Technologies* was first published in 1990, there have been a series of developments, most particularly in the area of personal communications tools. These range from new PCs to the potential creation of satellite-based personal communicators and new information highways. When viewed as a system, they could forever alter the way we produce, use, and relay information.

This second edition not only examines such new tools and their implications and applications but also describes the unchanged concepts originally presented in the first edition. These include the growth of digital systems, the implications of an information society, and fiber-optic applications.

As with the first edition, some of the technologies and applications presented in this book will undergo another series of changes by the time the book is published. This is an unavoidable consequence of writing about a volatile field that can change literally overnight.

On a more positive note, though, many of the *concepts* that underlie the technologies and their applications may remain essentially the same over the next few years. Even though new products may be introduced and enhancements adopted, fundamental principles may remain unaltered. Thus, by learning the concepts now, you may be able to work with the new applications well into the future.

Finally, getting down to basics, the following are some of the second edition's major changes and additions.

1. The chapters have been reorganized. Desktop video, for example, briefly introduced in the original Chapter 5, now has, in combination with multimedia, its own chapter (Chapter 8). This reorganization reflects the fields' potential impact in the overall communications market. Similar organizational changes have been implemented throughout the book.

2. Technological developments and their applications have been updated. In Chapter 3, for instance, new computer-based tools and their current and future applications are covered.

Discussions of relevant topics have also been expanded, including, in the same chapter, a more in-depth look at graphics programs, artificial intelligence, and potential health issues associated with computer use.

3. Emerging technologies and their applications, which didn't exist at the time of the first edition, are now included. The topics range from new, recordable optical disc systems to satellite developments.

4. Most of the illustrations are new, in line with the changes in the text. In one example, screen dumps highlight the discussion of different software applications.

5. The Additional Readings sections have been expanded. Topics now range from the technical to the legal implications of a given technology.

6. The convergence factor has been more fully developed. We look at the relationships between different fields and how developments in one area can influence another.

7. Perhaps the greatest change is in the coverage of nontechnical subjects. While the book still focuses on the technologies and their applications, more attention has been paid to their legal, social, political, and economic fallout. New topics include intellectual property, privacy, the First Amendment and the new technologies, the democratization of information, and aesthetics.

READERSHIP

The book is an appropriate text for communications technology courses in TV/Radio, Communication, and Journalism departments. It can also serve as a primer for graduate courses in the same departments and can be used as a supplementary text for management, public relations/advertising, and corporate communications courses.

The book may also prove useful for communications professionals who want to gain a broad overview of the field. This includes nonbroadcast practitioners, especially in light of the convergence of various technologies and their applications.

Finally, I hope the book will appeal to those interested in the communication revolution and its impact.

ACKNOWLEDGMENTS

I would like to express my gratitude to a number of individuals and institutions for their help in the production of this book.

As with the first edition, I want to acknowledge the pioneers, those individuals who helped launch the communication revolution itself.

I also want to thank my students for their willingness to serve as sounding boards for new material and for providing me with valuable insights. A number of individuals and institutions also graciously provided information, illustrations, and software, including Image-In, Inc., whose program was instrumental in producing the book's screen dumps. Staff members of the Federal Com-

munications Commission and the National Association of Broadcasters were also very helpful in answering questions, as was Mark Brender of ABC News.

I also wish to thank Phil Sutherland as well as the readers whose comments about the first edition and this book's proposal proved valuable and insightful. Ray "B.B." Ghirado and Jim Loomis also provided valuable suggestions, and I appreciate the efforts of Ayn Miralano, Martha Eckman, Dr. Holbrook, Dick Lanahan, H. Gogan, H. Rubin, M. Steele, and R. and L. Ortalano. Barbara Morgenstern, the first edition's co-author and contributor to this edition, similarly influenced the book's topics and overall organization.

Finally, the editorial and production staffs at Focal Press helped make this book a reality. I especially want to thank John Fuller, the production editor, Christopher Keane, the copy editor, and Sharon Falter and Marie Lee, this project's editors. Their enthusiasm, as well as their suggestions and sense of organization, have been crucial to this project's completion.

DISCLAIMER

Some of the companies and products mentioned herein are trademarks or service marks. Such usage of these terms does not imply endorsements or affiliations.

1 Communication in the Modern Age

A great deal has been written in recent years about the communication revolution—some of it realistic, some of it not. Authors predicted that videophones would permit us to see and hear our neighbors, every home would be wired for cable television, and satellites would make worldwide communications systems a reality. New communications lines would also support the mass distribution of electronic publications, the country would be crisscrossed by high-capacity fiber-optic lines, and we would witness the age of the personal computer.

Even though not all of these predictions have materialized as we approach the twenty-first century, many have. As we shall see, sophisticated satellites have created new communications nets, information ranging from voice to video is relayed in the form of light, and personal computers have forever altered the way we work.

We are clearly living in the age of the communication revolution. It is a time when a number of new and developing technologies, and, ultimately, equipment and systems, are greatly influencing the communications industry and society.

Similarly, other technologies and products, which may have existed for some time, are either influencing the market or are being used in new ways. For example, as discussed in Chapter 3, personal computers have been employed in different applications for a number of years. Yet the latest generation of hardware and software can now produce sophisticated multimedia presentations. In another example, optical-disc technology, examined in Chapter 4, experienced a boon in the mid- to late 1980s in the form of optical media that could store entire collections of sounds, photographs, and books on a handful of small discs.

Beyond these developments, other factors have contributed to the launching of the communication revolution: the mass production of integrated circuits and other chips that compose our contemporary communications equipment and systems and, as described in Chapter 2, the integration of digital technology.

In a complementary fashion, the phenomenal growth of the computer industry has played a major role in reshaping the communications field. The personal computer and supporting software have emerged as key tools, and pertinent applications are presented in almost every chapter. Thus, the communication revolution can be categorized as one in which new technologies, in combination with advances in preexisting technologies and fields, have brought about

FIGURE 1.1
MIT's Media Lab, one of the world's foremost research institutes, in fields ranging from computers to communications. (Courtesy of MIT Media Laboratory) [photo credit: Bea Bailey]

massive changes in the communications industry and the world around us.

But before we begin our exploration of these topics, several general characteristics of the communications field, and a series of technical principles and concepts, are presented in this chapter and the next. This background information will be referred to throughout the book.

BASIC CONCEPTS

Communications System—An Extended Definition

The term *communications system* is one such important concept. In essence, a communications system provides the means by which information, coded in signal form, can be transmitted or exchanged. If, for example, we decide to telephone our friend, the communications system would encompass the telephone receivers, the telephone line, which is the physical communications channel that plays a role in this exchange, and other assorted components. We currently employ a number of communications systems to exchange information and are developing ones that will enable us to exchange a greater volume of information, more rapidly, than at any other point in history.

In this book, the term *communications system* also encompasses a set of diverse hardware and software tools, including a computer graphics program, audio and video equipment, sophisticated electronic encyclopedias stored on optical media, and cellular telephones.

Consequently, the use of the term *communications system* is not limited to a description of different information-exchange systems. It has a broader definition, encompassing the communications tools we use, their applications, and the various implications that arise from the production, manipulation, and potential exchange of information.

Information

Another central concept is *information*. Information can be defined as a collection of symbols that, when combined, communicates a message or intelligence. When you write a note to a friend, for instance, specific letters of the alphabet and numbers are combined to convey your thoughts or ideas, a message that has meaning to both you and your friend. In this regard, the combination of the letters *c-a-r* is not just a collection of three letters. It represents the concept of a car or automobile.[1]

In our communications system, this information may be coded in a standardized form: in an electronic or electrical signal analogous to the coding of information by the printed letters, and ultimately words, in the written note.

Next, the information in the communications system can be transmitted via a telephone line, satellite, or other suitable communications channel or path. After it is received, it can be decoded or converted back into its original form.

This series of steps to exchange information follows the traditional model of a communications system devised by Claude Shannon and Warren Weaver. The system consists of an information source, a transmitter, the channel by which the information is delivered, a receiver, and a destination. Noise, another element, is described in the next chapter.

When coded in signal form, the information—a person's voice, a television image of the Grand Canyon—is also compatible with and can be used by a range of communications equipment. It can be transmitted, stored, and manipulated. Thus, a picture produced by a videocamera can be transmitted thousands of miles away, stored on videotape, or altered by a computer to produce special visual effects for a television program.

The communications system is quite flexible in how it represents information. Light produced by a laser can, for instance, convey information in a fiber-optic system, even though the information may have been previously coded in an electrical form. In another application, light produced by a laser is used to retrieve and represent information in optical storage media. These topics are covered in Chapters 5 and 4, respectively.

Those interested in reading a more detailed explanation of what constitutes information should see the books by Shannon and Weaver and by Rogers listed at the end of this chapter. They also cover the historical and technical elements of information theory, the mathematical theory of communication and information.

Information—An Extended Definition

The definition of information likewise extends beyond the typical information described in the context of television, radio, and communications. Information does not simply consist of television programs, computer data, telephone conversations, and music stored on CDs. Information is also a picture manipulated by a computer that highlights details of the human body. Information is a library of books that has been coded and stored on a handful of optical discs. Information is a series of facts that depicts the current state of the stock market, and that can then be accessed at home via a personal computer. These are only a few examples of information described in the book.

Information can be viewed as a commodity. In the traditional communications industry, television and radio programs are assigned a financial value that varies with a given show's ratings. The information generated by nontraditional tools can also be valuable.

A number of companies, as we shall see, are using the telephone system to sell information to users of personal computers. This information ranges from bibliographical material to demographic data about cities and towns across the country. In another setting, the television networks are purchasing photographs created by remote-sensing satellites. These pictures can reveal details of various regions of the Earth where important news events occur, such as the Chernobyl nuclear reactor incident.

Information can also be equated with power. Thanks to the widespread adoption of computers in all facets of society, a person who knows how to use a computer will potentially have access to a greater wealth of information than a nonuser. Such access

Figure 1–2
Mt. Saint Helens. Screen shot of a computer-generated image. You can now use PCs to re-create realistic as well as altered views of the Earth and even other worlds. (Software courtesy of Virtual Reality Laboratories; Vista Pro)

might give the user a political or economic edge.

The Information Society

Some technological developments, including the notion that information can be equated with power, have ultimately contributed to the creation of an *information society*. Even though this term has become something of a cliché, it is nevertheless quite accurate. Our society is driven by information, be it the latest business and financial news needed to keep abreast of a volatile world market or the creation of databases of information that can be accessed by computer owners.

For example, as described in Chapter 3, computers have been programmed to solve various problems, much like human experts. They can pinpoint a malfunctioning component in a complex piece of machinery. In this particular case, computers and the manipulation of information via expert-system software have altered the way we work.

The information age has also ushered in a variety of new job categories. Experts who can quickly access information from a collection of U.S. patents and other specialized databases are employed as independent consultants. A class of engineers, called knowledge engineers, is also emerging to support the expert systems just described.

In another example, the information age has transformed the economy of the United States. We are becoming a nation based on service rather than on manufacturing. Heavy industry, such as tubing and steel plants, has given way to the service industry. Hospitals, banks, and computer information companies all fall into the latter category; many use and process information as integral elements of the services they offer.

In a similar vein, many companies now produce the tools to sustain the information age, from computers to satellites to the sophisticated testing devices used in manufacturing these tools.

Finally, new highways have been constructed. But instead of carrying cars and trucks, they carry data quickly, accommodating a range of information and, potentially, delivering it to our homes.

IMPLICATIONS OF THE COMMUNICATION REVOLUTION

The new communications and the complementary computer and information technologies have profoundly affected our social structure. There is also a growing interdependence between technology, information, and society.

For example, the new communications technologies have raised a series of ethical questions, including, as detailed in Chapter 7, the use of computer-based scanning systems to duplicate another person's work. Using a scanner, a device that inputs graphic information to a computer, a user can alter a drawing or even use it in another publication without the original artist's approval or knowledge.

As presented in Chapter 3, some individuals are afraid that automation will accelerate the loss of jobs. This idea is typically linked with the belief that society is becoming increasingly dehumanized as the market is flooded with an ever-growing number of computers and computer-controlled systems.

A new social class is also forming in the United States. In the past, distinctions between social groups were influenced by economic and educational factors. These same forces are present today, but the distinctions have become even more pronounced through the growth of our information society. This is especially true for those who do not know how to use a computer or who lack information-management skills.

Look at the world around you. Banks are shifting to computer-based teller systems, and telephone lines are the keys to vast information resources. But unless you know how to tap this information, you will join the ranks of the information poor and be unable to compete, for example, for a higher-paying job that may require a certain level of computer proficiency. This state of affairs also reaches beyond the boundaries of the United States and influences whole nations that do not have the economic, political, or educational resources to participate in the information age.

This situation presents a paradox. More people now have access to a greater volume of information than at any other time in human history. Nevertheless, as we accelerate toward a world that will become even more dependent upon information, there are whole segments of society, on both national and international scales, that may not be able to partake of this information bonanza.

On the flip side, the communication revolution has numerous positive implications. The same computer systems that some people fear have enhanced the treatment of cancer patients. Doctors have drawn upon the experience and knowledge of experts in this field by using computer software. Other computer systems have made it possible for individuals with physical handicaps to communicate better with the world around them. It can also be argued that despite the unequal distribution of information in society, more people currently have greater access to a wider range of information, all at a lower cost.

Key Concepts

As indicated, the new technologies and their applications have raised a number of ques-

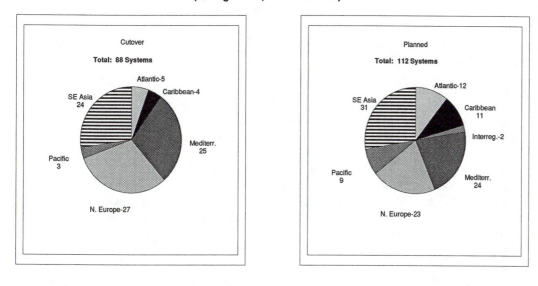

NUMBER OF FIBEROPTIC UNDERSEA SYSTEMS
INSTALLED TO DATE AND PLANNED BY REGION
(through 1991, and 1992-1997)

Source: 1992 Worldwide Summary of Fiberoptic Undersea Systems   © 1992 KMI Corporation, Newport, RI, USA

Figure 1–3
As described in later chapters, one of the important elements of the communication/information revolution is the fiber-optic line. This includes its use in the creation of high-capacity, undersea communications systems. (Courtesy of KMI Corp., Newport, R.I.)

tions and have various societal implications. While the focus of this book is on applications, important implications will also be discussed.

The latter topics cut across political, economic, legal, social, ethical, and aesthetic lines. They also play key roles in our information society and in fact have helped to shape it. As such, their current and potential impact merit discussion. This section serves as an introduction to some of the important implications.

Systems Approach. This book uses a systems approach.[2] While one can examine individual applications in isolation, to encompass other relevant areas to gauge key interrelationships and their impact on the overall communications system, one must broaden this outlook. Without such a broad perspective, we may be looking at an incomplete picture.

Our society's vulnerability to world events is one example that vividly highlights the importance of this approach and points out the interdependence between technology and our information society. In brief, the communications and information systems in

a wide geographical region, which could potentially cut across the United States, could be crippled and silenced by a single high-altitude nuclear explosion.[3]

This paralysis would be caused by a powerful *electromagnetic pulse* (EMP), a by-product of a nuclear detonation. Scientists have observed this phenomenon during weapons tests. They have subsequently used simulations to gauge the vulnerability of a wide range of electronic components, equipment, and systems that could be affected by this pulse, with the ultimate goal of developing protection mechanisms.

One step in this process was the initiation of a program by the Federal Emergency Management Agency to protect radio stations in the Emergency Broadcast System (EBS), among other communications centers. Theoretically, if a nuclear burst ever paralyzed our communications infrastructure, the protected EBS stations would be activated to supplement the damaged channels.

One potential problem with this scenario, though, is the receiving device. To pick up a broadcast, you may have to rely on and have access to a portable, battery-operated radio. It is one of the few modern electronic devices that may not be "readily damaged" by the EMP.[4]

But many of our radio receivers are part of stereo systems, which may be damaged. If not, there's a good chance the power plant that supplies the electricity to run the stereo would not function.

Thus, as a society that depends upon and is driven by information, the very tools that we use to create, manipulate, and deliver this information would be disrupted, if not rendered inoperable. The issue is multifaceted, and a broad systems perspective enables us to examine it from different angles and, more important, to cover all the key implications both inside and outside the communications field.

Finally, in keeping with this approach, it's ironic that vacuum tubes are relatively invulnerable to EMP effects. But in our contemporary communications system they have been universally replaced by solid-state chips, one of an EMP's favorite snacks. As our communications system was "modernized," it was made more vulnerable to world events and, potentially, to random acts of terrorism.

Convergence of Technologies. The communication revolution has accelerated the convergence of various technologies and, ultimately, their applications. As described in Chapter 8, a new generation of powerful PCs, when mated with the appropriate hardware and software, have created integrated desktop video systems.

These tools allow both individuals and organizations to create a variety of programs. When this capability is combined with other applications, such as desktop publishing, users are no longer just media consumers. We can also become producers and editors.

This newfound capability has, in turn, aesthetic implications, as discussed in Chapters 3, 7, and 8. Simply stated, the technical capacity to create a project does not supplant underlying aesthetic principles.

Intellectual Property. As discussed in Chapters 7 and 3, the new communications technologies raise a number of important intellectual-property issues, essentially the "rights of artists, authors, composers and designers of creative works,"[5] including the electronic copying of someone else's work as well as copyright and patent questions.

Another factor is information. In our current system, information is very malleable. It can be manipulated and stored easily. While this capability can be used for creative purposes, it also highlights a problem: the same family of tools that can create a graphic or other product can be used illegally to copy it. Since the tools—such as computers—are ubiquitous, it's almost impossible, as well as undesirable in most cases, to monitor their use.[6]

Copyright laws address these issues, and violators can be prosecuted. The difficulty, though, may lie in enforcing the laws in certain situations.

As outlined in Chapter 3, there is also a problem with recognizing intellectual property as actual property due to its sometimes intangible nature. One could, for example, spend a year or more developing a computer program that would possibly fit on just one floppy disk. On the surface, the program may not look like much, perhaps just a single disk you can hold in your hand. But in reality it may represent the "product of the creative intellect," that is, intellectual property, another form or type of property.[7]

The Democratization of Information. The communication revolution has introduced a number of tools that promote the free flow of information. You can use a desktop-publishing system, for example, to print a newsletter or even a book and subsequently place the document on the open market. The same system could work in league with fax machines and other communications tools to keep the world community apprised of fast-breaking political events. Both topics are discussed in Chapter 7.

This concept has been extended to other media. As outlined in Chapter 10, an electronic democracy could be sustained by holding electronic meetings between groups of people thousands of miles apart. A key element in this process is providing broad, affordable access to the system.

Planetary surface or data

Environmental image

NASA

Interactive surface visualization

Figure 1–4
As described in the Afterword, one of the more interesting applications is the development of virtual reality systems. This human–computer interface can support a range of applications. (Courtesy of NASA)

First Amendment Issues. The new communications technologies have also raised First Amendment questions. Our discussion will focus on defining the rights for various computer-based information services, as outlined in Chapter 10. For example, should an electronic service be treated as a distributor or as a publisher in First Amendment issues? Furthermore, what legal measures can be adopted to ensure that First Amendment rights and privileges are supported and nurtured in our new communications systems?

Privacy. The privacy question is one of the most pressing issues raised by the new technologies. As our communications systems become more sophisticated, so too does the potential capability to invade our privacy.

As examined in Chapter 10, concerns have been raised about the privacy of electronic mail, messages or mail relayed in electronic rather than print form. Another important issue is electronic eavesdropping. While new communications systems may be harder to tap, and special encryption schemes can be introduced, legislative forces could potentially make a system more vulnerable to electronic surveillance.

Economic Implications. As indicated in a previous section, the communication revolution has economic implications and in two specific cases can affect entire industries. These include, as outlined in Chapters 2 and 9, the computer and consumer electronics fields.

In the United States, another economic issue has to do with spectrum allocations. As described throughout the book, we use the spectrum as a means to relay information. In our society, this capability has an economic fallout.

OTHER ISSUES

Other issues have shaped and will continue to shape our information society. Will teleconferencing and telecommuting, two applications covered in Chapter 11, reduce human communication to a machine-dominated format? Do computers dehumanize the creative process? Or do they in fact extend our creative capabilities and enable us to transform an idea or vision into a concrete reality, an actual product? This topic is discussed in Chapters 3 and 8.

CONCLUSION

Chapter 1 has introduced the universe of new technologies, presenting terms that will be used throughout the book and pointing out some of the issues facing the communications industry and society as a whole.

The rest of the chapters will follow the pattern established by Chapter 1. A brief overview of the pertinent technology provides the technical underpinnings that will help us examine and, inevitably, manage a technology and its applications. This knowledge may also prove valuable when deciding if a technology could be used for a specific application, and possibly for developing applications yet undiscovered.

Besides this technical brief and the discussion of applications, which is this book's focus, we will present some of the implications—ranging from legal to social to economic—raised by the integration of technologies and their applications.

REFERENCES/NOTES

1. Frederick Williams, *The New Communications* (Belmont, Calif.: Wadsworth Publishing Co., 1984), 9.
2. For more information, see Ervin Laszlo, *The Systems View of the World* (New York: George Braziller, Inc., 1972).
3. A single nuclear explosion could theoretically blanket the United States and parts of Canada and Mexico. It could disrupt the countries' communications infrastructure even if the weapon did not cause any direct damage.
4. Samuel Glasstone and Philip J. Dolan, *The Effects of Nuclear Weapons* (Washington, D.C.: U.S. Government Printing Office, 1977), 521.
5. Westlaw, Intellectual Property Database, screen dump, December 1, 1992.
6. Monitoring, for example, could lead to censorship.
7. "Patents: Protecting Intellectual Property," *OE Reports* 95 (November 1991): 1.

ADDITIONAL READINGS

Carter, A. H., and members of the Electrical Protection Department. *EMP Engineering and Design Principles.* Whippany, N.J.: Bell Telephone Laboratories, Inc., Technical Publication Department, 1984. And, Pierce, John R., Chairman, Committee on Electromagnetic Pulse Environment; Energy Engineering Board; Committee on Engineering and Technical Systems; National Research Council. *Evaluation of Methodologies for Estimating Vulnerability to Electromagnetic Pulse Effects.* Washington, D.C.: National Academy Press, 1984. Both publications touch upon different elements of the EMP issue. The first publication provides a good overview of the creation and technical implications of an EMP as well as different equipment and protection schemes, among other topics. The second publication covers a wide range of subjects, including the role of statistics in trying to predict potential equipment and system failures and how to protect potentially susceptible systems.

Rogers, Everett M. *Communications Technology.* New York: The Free Press, 1986. An excellent reference and resource book, it examines the history of communications science, the social impacts of communications technologies, and new research methods for studying the technologies.

Shannon, Claude, and Warren Weaver. *The Mathematical Theory of Communication.* Urbana, Ill.: University of Illinois Press, 1949. *The* book about information theory. Also see Bharath, Ramachandran, "Information Theory," *Byte* 12 (December 1987): 291–298, for a discussion of information theory; and Tufte, Edward, *Envisioning Information* (Cheshire, Conn.: Graphics Press), for an in-depth examination of the visual representation of information.

The following publications can be used to provide an overview of some of the issues facing our society through the introduction of communications and related technologies. Other publications, which may similarly cover these and related issues, are listed in appropriate chapters.

Cleveland, Harlan. "How Can 'Intellectual Property'

Be Protected." *Change* 21 (May/June 1989): 10–11.

Fisher, Francis Drummer. "The Electronic Lumberyard and Builders' Rights." *Change* 21 (May/June 1989): 13–21.

"Get On Board for the Future." *Broadcasting,* November 30, 1987, 58–70.

Gross, Lynne Schafer. *The New Television Technologies.* Dubuque, Iowa: Wm. C. Brown Publishers, 1990.

Head, Sidney W., and Christopher H. Sterling. *Broadcasting in America.* Boston: Houghton Mifflin Co., 1990.

Hoffman, Gary M. *Curbing International Piracy of Intellectual Property.* Washington, D.C.: The Annenberg Washington Program, 1989.

Inglis, Andrew F. *Behind the Tube.* Boston: Focal Press, 1990.

McPhail, Thomas L., and Brenda M. McPhail. *2001: A Strategic Forecast.* McPhail Research Group, 1990.

Middleton, Kent R., and Bill F. Chamberlin. "Intellectual Property." In *The Law of Public Communication.* New York: Longman Publishing Group, 1991, 240–77.

Murphy, John W., and John T. Pardeck, eds. *Technology and Human Productivity.* New York: Quorum Books, 1986.

National Telecommunications and Information Administration. *Telecommunications in the Age of Information.* NTIA Special Publication 91-26, October 1991.

Speser, Philip, and John D. Hill. "Technology Transfer: One Key to a Competitive Edge." *Laser Focus* 22 (June 1986): 18–20.

Truxal, John G. *The Age of Electronic Messages.* New York: McGraw-Hill Publishing Company, 1990.

U.S. Congress, Office of Technology Assessment. *Critical Connections: Communication for the Future.* Washington, D.C.: U.S. Government Printing Office, 1990.

U.S. Congress. Public Law 100-568; October 31, 1988, 100th Congress, "Berne Convention Implementation Act of 1988," *Statutes at Large.* Washington, D.C.: U.S. Government Printing Office, 1990.

GLOSSARY

Communications system: The means by which information, coded in signal form, can be transmitted or exchanged. In the context of this book, the communications system also encompasses the communications tools we use, their applications, and the various social, ethical, and economic implications that arise from the production, manipulation, and potential exchange of information.

Electromagnetic pulse (EMP): A by-product of a nuclear explosion; a brief but intense burst of electromagnetic energy that can disrupt and destroy integrated circuits and related components.

Information: In the context of this book, a collection of symbols that, when combined, communicates a message or intelligence.

Information society: In effect, a society that has shifted from a heavy, industrial base to one that is driven by the production, manipulation, and exchange of information. In this setting, information can be viewed as a social, economic, and a political force.

Intellectual property: As defined in the chapter, the rights of artists, authors, and designers of creative works; the products of the creative intellect.

2 Technical Foundations of Modern Communication

This chapter introduces the various technical elements that are the foundations of our modern communications system. The chapter is also a continuation of Chapter 1 but concentrates more on technical information and on an important development in the communications industry, digital communication.

The chapter concludes with a discussion of the importance of technical standards. A standard can help promote the growth of a communications technology and industry. If a standard is not adopted officially or unofficially, the industry's growth could be hampered. This situation had, in fact, curtailed the growth of the teletext industry and had an impact on the development of a new improved television system. These topics are discussed in Chapters 10 and 9, respectively.

BASIC CONCEPTS

The Transducer
A transducer is a device that converts one form of energy into another form of energy. When a person talks into a microphone, the device converts his or her voice—sound or acoustical energy—into electrical energy, or in more familiar terms, an electrical signal. A speaker, also a transducer, connected at the other end of the line can reconvert the electrical signal—an analog signal—back into the person's voice.

Certain transducers can be considered extensions of our physical senses since they can convert what we say, hear, or see into signals that can be processed, stored, and eventually transmitted.[1] Thus, a videocamera is analogous to a human eye. The camera scans brightness levels, patterns of light, and changes the scene into an analog electrical signal. Once the signal is transmitted, a television set can reconvert that signal back into a representation of the original scene.

To carry this concept one step further, we can consider "spoken words," sound waves, to be a natural form of information, which our senses "perceive."[2] A transducer converts this category of natural information, among others, into an electrical representation. The transducer in this situation acts as a link between our communications system and the natural world, made up of natural analog information or signals, that is, the sound waves.

The Characteristics of a Signal
If the operation of a transducer, such as a microphone, were visible, we would see what appears to be a series of waves traveling through the line connected to the microphone. The sequence of waves, the electrical representation of the person's voice, has a number of distinct characteristics. Two that are pertinent to our discussion are *amplitude* and *frequency*.

The amplitude is the height of a wave and, in our example, corresponds to the signal's strength or the volume of the person's voice. The frequency, the pitch of the voice, can be defined as the number of waves that pass a specific point in one second. If a single wave passes the point in a second, the signal is said to have a frequency of one cycle per second (cps). If a thousand waves pass the same point in one second, the signal's frequency is 1,000 cps.

The signal's frequency, the cps, is usually expressed in Hertz (Hz), after Heinrich Hertz, one of the pioneers whose work made it possible for us to use the electromagnetic spectrum, a keystone of our communications system. Hence, a frequency of 10 cps is written as 10 Hz; higher frequencies are expressed in kilohertz (kHz) for every thousand cycles per second, megahertz (MHz) for every million cycles per second, and gigahertz (GHz) for every billion cycles per second.

Analog Signal

As stated by Simon Haykin in his book *Communication Systems*, "Analog signals arise when a physical waveform such as an acoustic or light wave is converted into an electrical signal."[3] Many communications devices, such as telephones, videocameras, and microphones, are analog devices that create and process analog signals.

In the case of a microphone, this signal, an electrical representation of the person's voice, is said to be continuous in amplitude and

time. The amplitude, for example, can essentially assume an infinite number of variations or levels within the operational bounds of the communications system.

The signal is an "analogue" (analog), that is, representative of the original sound waves. As the sound waves change, so too do the signal characteristics in a corresponding fashion.

Digital Signal

A digital signal is "a non-continuous stream of on/off pulses. A digital signal represents intelligence by a code consisting of the sequence of discrete on or off states."[4]

A digital system uses a sequence of numbers to represent information, and unlike an analog signal, a digital signal is noncontinuous. As discussed later in this chapter, an analog signal can be converted into a digital signal through an analog-to-digital conversion process.

The Electromagnetic Spectrum

The electromagnetic spectrum is the entire collection of frequencies of electromagnetic radiation, ranging from radio waves to X rays to cosmic waves. Infrared and visible light, radio waves, and microwaves are all well-known elements and forms of electromagnetic energy that compose the electromagnetic spectrum.

We tap into the spectrum with our communications devices and use the electromagnetic energy as a communications tool—a means to relay both analog and digital information. For example, radio stations employ the radio frequency range of the spectrum for transmission purposes.

Spectrum space—allocated, for example, to television and radio stations for over-the-air operations—can also be viewed as a commodity in our information society. The electromagnetic spectrum can be used to generate income, and like oil, gas, and other valuable natural resources, it is scarce. In this light, the spectrum has an intrinsic worth and monetary value.

The issue of spectrum scarcity, which has also been used as a basis for regulating the electronic media, is a reflection of our com-

Figure 2–1
Electromagnetic spectrum. (Courtesy of the Earth Observation Satellite Company, Lanham, Md.)

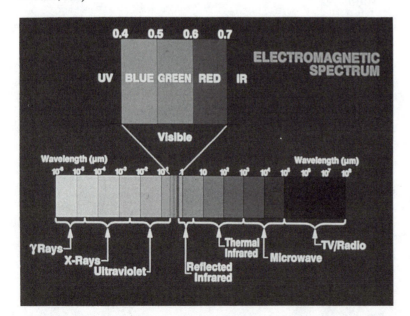

munications system. We can't use all the spectrum for communications purposes, and the portions that are available are divided nationally and internationally. In the United States, allocations are controlled by the National Telecommunications and Information Administration (NTIA) and the Federal Communications Commission (FCC) for governmental and nongovernmental interests, respectively.

New and emerging technologies, including the personal communications networks discussed in Chapter 11, are exacerbating such scarcity. The demand for spectrum space is increasing, and the government is hard pressed to provide users with the necessary frequency assignments. In fact, when a small portion of the spectrum opened up for mobile communications services in 1991, the FCC received approximately 100,000 applications for this space in a three-week period.[5]

Because the spectrum is such a valuable resource, various governmental agencies have called for new spectrum-management and allocation policies, including imposing fees for spectrum usage and auctioning new allocations.

Proponents of such policies claim that auctioning would generate additional revenues; mechanisms could be initiated to ensure that the public-interest mandate, an important regulatory ideal, would still be met.[6] Opponents, on the other hand, have claimed that auctioning conflicts with the notion that the spectrum is a public resource, not private property that can be sold and bought.

While the debate has heated up recently, it is not new. In 1964, for example, Ayn Rand wrote that "the airwaves should be turned over to private ownership. The only way to do it now is to sell radio and television frequencies to the highest bidders (by an objectively defined, open, impartial process)."[7] Rand, an author, philosopher, and advocate of capitalism, indicated that private rather than public ownership of the airwaves would have protected the electronic media from government regulation and would have ensured an open, competitive, and free marketplace.

In another twist, some individuals have called for switching services: television should be delivered by cable, thus freeing the spectrum for wireless communications services. The cable, owned by cable-television or potentially telephone companies (as discussed in Chapter 5), would most likely be part of a fiber-optic system. This proposal, also known as the "Negroponte Switch," after Nicholas Negroponte, the director of the Massachusetts Institute of Technology's (MIT's) Media Lab, may solve an element of the spectrum scarcity and management problem.[8]

This type of switch, though, could not take place overnight. A question also arises: Over-the-air television is free. Will consumers still receive at least a portion of television programming for free if it is relegated to a cable-based delivery system?

Modulation

The term *modulation* can be defined as the process by which information is superimposed upon a carrier. Modulation is associated with, but not limited to, the communications systems most familiar to us, including AM and FM radio stations. When information, such as a disk jockey's voice, is transduced or converted into an electrical signal by a microphone, it can be superimposed or impressed upon a higher-frequency carrier signal for transmission purposes. In this regard, a physical characteristic of the carrier is altered to convey, or to act as a vehicle to carry, the information. Prior to this time, the carrier wave did not convey any intelligence.

Next, after the signal from our radio station is received, the original information can be stripped, in a sense, from the carrier so we can hear the disk jockey's voice over the radio's speaker. Both analog and digital information can be relayed through such modulation schemes.

Bandwidth

For the purpose of our discussion, a communications channel's *bandwidth*, its capacity, dictates the range of frequencies, and to all intents, the categories and volume of information the channel can accommodate in a given time period.[9]

There is a relationship between a signal's

frequency and its information-carrying capability. As the frequency increases, so too does the capacity to carry information. The signal must then be relayed on a channel wide enough to accommodate the greater volume of information.

A television broadcast signal, for example, has a higher bandwidth requirement than either a radio or a telephone signal due to its greater (unit time) information content. Consequently, under normal operating conditions, a voice-grade telephone line cannot carry a conventional television signal.

Noise

During the exchange of information, *noise* will be introduced into the communications system, adversely affecting the quality of the transmission. If the noise is too severe, it may distort the relayed signal or render it unintelligible, and the information may not be successfully exchanged. For example, if the snow on a television screen—the noise distorting the relayed signal—is very severe, it may become impossible to view the picture.

Noise can be internal, introduced by the communications equipment itself, or external, originating from outside sources. The noise may be machine-generated or natural.

Lightning is one such natural source of noise. It is fairly common and may be manifested in the form of static that disrupts a radio transmission. The sun is another natural noise source, creating interference through solar disturbances and other phenomena. Machine-generated noise, on the other hand, may derive from the electric motors in vacuum cleaners or large appliances, for example.

When discussing noise relative to a communications system, the term *signal-to-noise ratio* is frequently encountered. The signal-to-noise ratio is a power ratio, that of the power or strength of the signal versus the noise. For information to be successfully relayed, the noise must not exceed a certain level. If it does, the noise will disrupt the exchange to a degree dependent upon its strength, and it will have a direct impact upon a communications channel's quality and transmission capabilities.

DIGITAL TECHNOLOGY

Now that we've examined some of the underlying principles of our communications system, it's possible to investigate the actual technologies that helped launch the communication revolution. One such area is digital technology and its applications.

Digital technology has played a vital role in the development of new communications lines, information-manipulation techniques, and equipment. Preexisting communications channels and devices have also been affected by this technology. But before we can explore this digital world, the concept of what is meant by "digital" must be investigated.

As already stated, many of the communications devices we use, such as videocameras and microphones, create and process analog rather than digital signals. Due to their distinct characteristics, digital and analog signals, and ultimately equipment and systems, are generally not mutually compatible. This situation mandates the adoption of an analog-to-digital and a digital-to-analog conversion process in order to use a mixed bag of analog and digital equipment, and what are considered to be analog and digital communications channels, in the overall communications system. The two-way link is crucial to the integration of digital technology in the contemporary communications structure, which is, to an extent, based upon an analog standard. This is an important consideration since digital technology has distinct advantages over its analog counterpart.

Different categories of information can be represented and consequently transmitted over the appropriate channels, in both analog and digital forms. A standard voice or audio signal, for instance, is analog in nature. Yet, through an analog-to-digital conversion process, the analog information can be converted into a digital representation and subsequently relayed.

Analog-to-Digital Conversion

In an analog-to-digital conversion, the analog signal is converted into a digital signal.

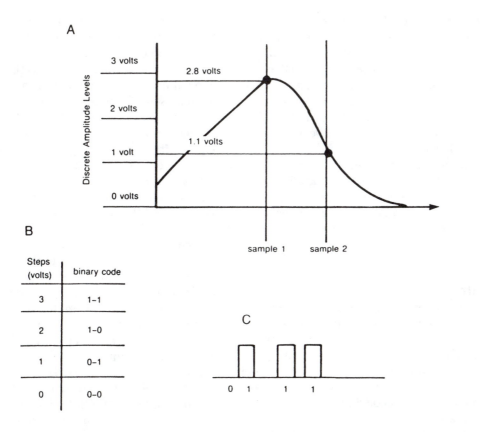

A

B

C

Figure 2–2
*A PCM operation. The
diagram has been sim-
plified for illustration
purposes. The two sam-
pling points have been
indicated by samples 1
and 2 (A). Sample 1 is
2.8 volts and is assigned
to the nearest step, 3
volts. Sample 2, 1.1
volts, is assigned to the
nearest step, 1 volt.
Based on the chart (B),
the samples are coded
as 1-1 and 0-1, respec-
tively. The result is the
PCM signal (C).*

Binary language is the heart of digital communication; it uses two numbers, 1 and 0,
arranged in different codes, to exchange information. The numbers 1 and 0 are called
bits, from the words *binary digits,* and represent the smallest pieces of information in a
digital system. They are also the basic building blocks for a widely used digital information system, *Pulse Code Modulation* (PCM). A
simplified explanation of the general principles that govern this process follows.

PCM is a coding method by which an
analog signal can be converted into a digital
representation, a digital signal.[10] PCM information consists of two states, either the presence or the absence of a pulse, which can also
be expressed as "on" or 1 and "off" or 0.

When the analog signal is actually digitized, it is sampled at specific time intervals.
Rather than converting the entire analog
signal into a digital format, it is sampled or
segmented, and only specific parts of the
signal are examined and converted through
this procedure. Enough samples are taken,

however, to obtain a sufficiently accurate
representation of the original signal.

The samples are then compared to, for
illustration purposes, a preset scale composed of a finite number of steps. The steps
represent different values or amplitudes that
the original analog signal could assume.[11] A
sample is assigned, in a sense, to the step that
matches or is closest to the sample's amplitude. Each step, in turn, corresponds to a
unique word composed of binary digits (for
example, 11 or 01).

A sample is then coded and represented by
the appropriate word. The word can be relayed as on and off pulses, and when the
information reaches the end of the transmission line, it is detected by the receiver. Ultimately, a value corresponding to the original
analog signal at the sampling point has been
transmitted since the word represents a
known quantity. (For more specific details,
please see the books and articles at the end
of this chapter.[12])

Finally, for the purpose of our discussion,

the analog signal is converted into a digital format by an *analog-to-digital converter* (ADC). Once the coded information is relayed, it can be reconverted into the original analog signal by a *digital-to-analog converter* (DAC) to make the signal compatible, once again, with analog equipment and systems. Hence, ADCs and DACs act as bridges between the analog and digital worlds.

Morse Code functions in a similar fashion: information, a message, is coded, in this case as a series of dots and dashes. Following its relay, the original message can be reconstructed by an operator since the code represents a known value.

Steps

In many communications systems, when the analog signal is digitized, it is generally coded as either seven- or eight-bit words. There is a direct relationship between the number of bits in a word and the number of steps: as bits are added, the number of steps that can represent the analog signal increases in kind.

If an eight-bit-per-word code is used, essentially 256 levels of the signal's strength can be represented. This number is derived from the binary system, in which an eight-bit code is the equivalent of two to the eighth power ($2 \times 2 \times 2 \times 2 \times 2 \times 2 \times 2 \times 2$). In this system, a different combination of eight 1s and/or 0s represents each step. This number is significant since an increase in the number of steps can lead to a more precise representation of the original signal. In addition, certain communications systems and data require a large number of steps, and consequently levels, for an accurate representation. Nevertheless, they are limited in comparison with the enormous range of levels the analog signal can potentially assume.

Sampling and Frequency

In a PCM system, the *sampling rate,* which is the number of times the analog signal is sampled per second, is another vital element in the reproduction process. The primary goal of sampling is to reflect accurately the original signal through a finite number of individual samples.

A survey conducted across the United States to predict the outcome of a presidential election can serve as an analogy. A sample of individual voters represents the voting population as the samples of the analog signal represent the original signal.

The sampling rate employed in an analog-to-digital conversion is based upon a signal's highest frequency in a given communications system. If a signal is sampled at a rate that is at least twice its highest frequency, the analog signal will be accurately represented. This is called the Nyquist rate.

For a standard voice-grade telephone line, the communications channel carries only frequencies below 4 kHz, and the sampling rate for a line is 8,000 samples per second. Enough samples are generated to reproduce the analog signal and, in this instance, a person's voice. It should also be noted that higher sampling rates can be employed in different communications systems.

After the analog signal is sampled and coded, the final transmission can be composed of millions of bits. In one format, the telephone industry employs an eight-bit-per-word code when the analog signal is digitized, and a telephone conversation can be transmitted at the rate of 64,000 bits per second (64 kilobits/second). Other digitized analog signals similarly generate high bit rates.

This volume of information poses a problem for some communications systems, since they may not have the channel capacity to transmit the information, and special lines that can accommodate this information flow must be used. One such line is AT&T's T1 carrier, which can transmit 24 digitally coded telephone channels at a rate of 1.544 million bits per second (1.544 megabits/second). This data stream is also composed of bits that ensure the integrity of the data and satisfy other technical and operational parameters.

The T1 line is one of the workhorse channels of the communications industry, and transmissions take place both inside and outside of the telephone system. It is also a flexible standard and can integrate voice and data, so one communications channel can carry different types of information. The T1 line has the additional capacity to accommo-

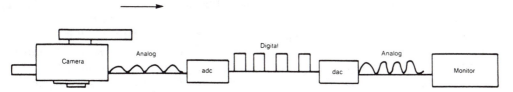

Analog to Digital Conversion

Figure 2–3
An ADC and DAC operation. An analog signal is converted into a digital domain via the ADC, and back into analog with the DAC.

date more specialized needs, such as relaying video information, when necessary.[13]

ADVANTAGES OF A DIGITAL COMMUNICATIONS SYSTEM

Now that digital communication has been described, we can examine some of the major advantages of digital communications equipment and systems. These features have provided and will continue to provide the impetus for their continued use and growth.

Computer Compatibility

Once a signal is digitized, it can be processed by a computer. The capability to manipulate digitally coded information, such as pictures produced by a videocamera, is central to the video and audio production, medical, and desktop-publishing industries, among others.

In the case of a videocamera, the video signal, an electrical representation of the light and dark variations, the brightness levels of the scene the camera is shooting, is converted into a digital form. The digital information represents *picture elements* (pixels), a number of small points or dots that actually make up the picture.[14]

In a black-and-white system, a pixel assumes a specific level or shade of the gray scale. The gray scale refers to a series or range of gray shades, as well as the colors black and white, that compose and reproduce the scene.

A pixel is represented by a binary word that is equivalent to a level of this scale. The level, in turn, is delineated by the brightness value of the corresponding section of the original scene now represented by the pixel.

The actual number of gray levels is determined by the number of bits assigned to a word. This characteristic is important for the accurate reproduction of the original scene. If too few bits are used, only a limited number of gray shades will be supported.

Once the picture has been digitized, a computer with the appropriate software can manipulate this image data. A special effect can be created, or in various science disciplines,

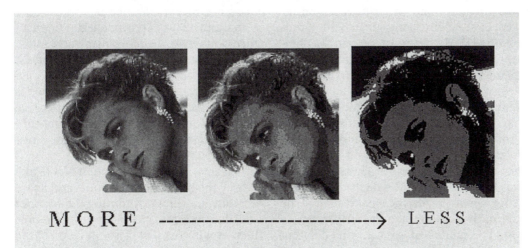

MORE ----------------------------> LESS

Figure 2–4
The visual effect of reducing the number of gray levels in an image. (Software courtesy of Image-In, Inc.; Image-In Color)

picture qualities and defects can be respectively enhanced and corrected. For example, an unfocused picture can be corrected to a varying degree to produce a sharper image.

Information in digital form, including picture information, also lends itself to mass storage. After the analog signal is digitized, the information can be recorded on a variety of media, including hard drives and optical discs, both of which are discussed in later chapters.

Finally, the digitization of a videocamera's signal is not limited to black-and-white systems. Color operations are also popular, but due to the more complex nature of color information, the configuration itself is much more complex.

Multiplexing

Digital and analog signals can be multiplexed. Multiplexing is a process in which multiple signals are relayed on a single communications line. The signals share the same line for transmission purposes, and thus, fewer lines have to be constructed and inevitably serviced and maintained.

Two advantages offered by both analog and digital multiplexing are, in fact, the cost- and labor-saving properties. In an analog-based configuration, frequency division multiplexing (FDM), a communications line is divided into a number of separate and smaller channels, each with its own unique frequency. For a telephone system, the various telephone signals have been assigned to these separate channels, processed, and relayed. In another system, time division multiplexing (TDM), time, not frequency assignments, separates the different signals. A TDM scheme can make for a very effective relay when conducted in the digital domain; an example of this operation is the T1 line as employed by the telephone industry.

The T1 system is digital in form, but the typical telephone receiver and local telephone system work with analog information. This necessitates the digitization of the analog signal, the electrical equivalent of the voice of the person using the telephone.

Briefly, groups of signals, the different conversations, are divided into smaller pieces for transmission. Each signal is relayed a piece at a time, in specific time slots.

In terms of the T1 carrier in this particular configuration, digitized information from 24 channels is transmitted one after the other on the line. The information is actually organized in a grouping, called a frame. Following the transmission, the 24 channels are separated.[15]

As a PCM-based operation, this digital multiplexing system is very fast, efficient, and clean, with regard to the signal's quality. As a digital operation, it can also be readily controlled and monitored by computers with all of the attending advantages, namely, computer accuracy and speed.

The Integrity of the Data When Transmitted

When digital signals are transmitted over long distances, the integrity of the data is preserved even after a signal has passed through a number of repeater stations. A repeater is a device that strengthens or boosts a signal as it travels along its transmission path.

When an analog signal passes through a repeater, noise that adversely affects the signal may be introduced in the transmission. This problem can potentially be multiplied by the number of times the signal is boosted. Noise can accumulate, and the signal's quality can progressively deteriorate.

Digital systems are not affected by this operation. Instead of boosting a signal, the pulses are regenerated, or new pulses are created and relayed at each repeater site. This process makes a digital transmission much hardier since a new signal, rather than a boosted one that may be adversely affected by noise, is relayed. A digital transmission is also less susceptible to noise and interference in general, and this quality further contributes to a digital system's superior transmission capability.

This factor led to the adoption of a digital relay by the National Aeronautics and Space Administration (NASA) during the Mariner 4 mission to Mars, in 1964. Mariner 4 provided us with the first close-up views of the planet, and the digital relay helped preserve the integrity of the information as it traveled through millions of miles of space.[16]

The Flexibility of Digital Communications Systems

Digital systems are flexible communications channels that can carry information ranging from computer data to digitized voice and video signals. In an all-digital environment, analog signals, such as those produced by telephones and videocameras, would be digitized, while computer data, already in digital form, would be accommodated without this type of processing. A digital configuration would also eliminate the use of a conventional modem, even though a special adapter may be required to connect a computer, for example, to the line.

The modem, an acronym of the words *modulator* and *demodulator*, is widely used in the current communications system to relay computer data over a standard voice-grade telephone line. The modem converts a computer's digital information into a form that complies with the technical characteristics of the line, and information is relayed as a series of audio frequencies or tones through a modulation technique. At the other end of the line, the information is converted back into its original form by a second modem.

Data relays in the consumer and business markets are typically conducted at 2400, 4800, and 9600 bits per second, or even higher through data compression.[17] The 9600 rate is popular in the business sector and is becoming more popular with computer hobbyists and other users due to its faster relay capabilities: as the modem's speed increases, the telephone connect time with the other computer decreases. This makes the relay more efficient and, potentially, more cost-effective.

Digital communications systems, on the other hand, can support faster relays and benefit from a digital transmission's superior technical characteristics and performance. A digital system can also integrate a wide range of information on a single line, instead of using separate communications channels.

One of the most promising digital platforms or networks is the *Integrated Services Digital Network* (ISDN).

The ISDN "can be thought of as a huge information pipe, capable of providing all forms of communications and information (voice, data, image, signaling). . . . It's an information utility accessible from a wall outlet . . . with a variety of devices able to be simply plugged and unplugged."[18]

There are two ISDN-based services, Basic and Primary, and they support the following transmission schemes:

Basic Rate Interface (2B + D): two 64 kilobits/second B channels for information (e.g., voice) and one 16 kilobits/second D channel for signalling and control information.
Primary Rate Interface (23B + D): 23 64 kilobits/second B channels and one 64 kilobits/second D channel. For Europe, the Primary Rate Interface is 30B + D.[19]

The ISDN could revolutionize the way we communicate. We could exchange pictures, computer graphics, and data with the same regularity and ease with which we use a telephone for conversations. Developments of this nature will continue to reshape the communications industry. In the case of the ISDN and other digital standards, we will be able to select the proper medium for the form of communication we are undertaking rather than being bound to a voice-only operation.

An architect talking on the telephone could, for example, send a picture of a blueprint. This capability would add another dimension to the communication process since important visual information could also be readily relayed.

Prior to this time, the architect may have been limited to describing the plans over the telephone. Or, a special communications configuration, a teleconferencing outfit, could have been assembled. As described in another chapter, this alternative may mandate the use of special equipment and either a second telephone line or a high-speed communications channel to relay the visual information. An ISDN would simplify this operation. Additional communications lines would not have to be used or even installed, and the link between the sites would be seamlessly and easily established.

This development would help make the technology *transparent*. In a similar way that we may drive a car or use a telephone without thinking about the technologies behind these products, the ISDN may make it possible to relay different types of information without having to think about the actual process. We do not have to be concerned with how the information is sent to another location since it is automatically handled for us.

In sum, the ISDN could provide users with a dynamic communications tool that would work in conjunction with preexisting and newly constructed lines. Furthermore, rather than locking us into a system that cannot adopt new technologies and equipment that would enhance the communication process, the ISDN should be flexible enough to incorporate these advances.

There are some problems, though, associated with the ISDN platform as of this writing. The first is the uncertainty of standards, the technical parameters that govern a system's facilities and operation. For example, before the early 1990s and the adoption of the National ISDN-1 standard, equipment from different manufacturers was not necessarily compatible.

A second complaint focuses on the ISDN's transmission rate. Some individuals believe it will not meet future transmission needs, especially in light of the enormous volume of information computers, related peripherals, and high-definition video systems will generate. But this problem may be solved by *Broadband ISDN* (B-ISDN), an enhanced version of the original ISDN concept, now called *Narrowband ISDN* (N-ISDN).

Digital-compression techniques, some of which are described in Chapter 8, may also make it possible to use N-ISDN lines for applications it was first thought they could not support. Other high-speed communications links are likewise planned.

A third problem is availability. Digital transmission systems, such as the ISDN, are not universal. There may be certain geographical restrictions, in respect to easily establishing a link between a customer and a telephone company's central office. The office must also be suitably equipped for this operation.

Consequently, it may be a number of years before certain customers are able to take advantage of an ISDN-based service, or other configuration, as the telephone industry upgrades the current physical plant. When completed, even the local loop, essentially the line between your home or business and a central office, should be able to support various digital services.

This initial, slow integration of digital technology in the physical plant has not been mirrored in other countries. According to an NTIA report, Japan and various European countries are leading the United States in the development of national ISDN systems.[20]

The modernization of a nation's telecommunication infrastructure, be it with the ISDN or another digital system, is vital. As stated, information is a commodity in our world. The country or countries that can create an efficient information-exchange system on national and international levels will have a decided advantage in competing in the world market. There are also political ramifications, as discussed in Chapter 10.

Cost-Effectiveness

As digital equipment is mass-produced and the manufacturing costs are reduced, digital communications systems become increasingly cost-efficient to build and maintain. Digital equipment and communications systems are also more stable and require less maintenance than comparable analog configurations.

This positive feature should enable an organization to function with a smaller maintenance department, and it could reduce its expenditure in this area. More important, the communications system should remain operational with fewer regularly scheduled maintenance calls.

This superior stability and durability is partly due to the integrated circuit (IC). An IC is a semiconductor, a solid-state device. It is a single chip or unit that may contain the equivalent of well over 100,000 separate transistors, diodes, and resistors, the fundamental units of electronic communications equipment. In essence, the IC is one of the much-talked-about chips that have helped launch the communication revolution.

The creation of such a chip had and still has an enormous impact upon the communications industry. Instead of wiring thousands of individual components while building a piece of equipment, a single chip is used. If the equipment malfunctions, only one chip may have to be replaced, and it's easier to isolate component failures, thus further reducing the necessary maintenance and time expenditures.

In addition, ICs have made it possible to manufacture compact equipment, since a limited number of chips are used rather than the multiple components. This size differential has also contributed to declining product costs. If less raw material is now used to make a piece of equipment, it costs less to build, and eventually, it should sell at a lower price than previous models.

But ICs are not without their own set of problems. The initial development cost may be high, and until the chips are mass-produced, the early units can be expensive. The chips also tend to be vulnerable to static charges and power surges. If a chip is exposed to static electricity, for instance, its operational capabilities may be temporarily or permanently disrupted.

Political issues may also have an impact on the cost-effectiveness of digital systems. In one example, the price of computer memory chips soared during the 1980s. The Reagan administration tried to protect American chip manufacturers from Japanese competitors, and a short-term outcome of this political maneuvering was a jump in chip prices. At one point, a single, common memory chip was more expensive than an ounce of silver.

The 1990s witnessed a similar situation. Small, portable computers fitted with liquid crystal display (LCD) panels became very popular. A group of American manufacturers subsequently charged the Japanese with dumping inexpensive panels on the market, and a tariff was levied on imports of active-matrix flat-panel displays. The tariff was im-

Figure 2–5
One of the fallouts of the communication revolution: cost-effective and powerful PC-based video editing systems. These systems are described in Chapter 9. (Courtesy of Avid Technology, Inc.; Media Composer)

posed to help support American manufacturers in this area.

While the tariff could have helped American manufacturers in the long term, there was a short-term consequence. Imported computers already fitted with the displays were exempt from the tariff. Consequently, American computer companies threatened to move their manufacturing facilities offshore to avoid the additional fee.[21]

DISADVANTAGES OF A DIGITAL COMMUNICATIONS SYSTEM

With any technology, there are disadvantages as well as advantages. Digital technology is no exception.

Greater Channel Demands

Once analog information is digitized, the large volume of bits generated during the operation mandates the use of a communications system with a greater channel capacity, and this factor adds to the transmission's overall cost. There are ways, however, to compress the digital signal's bandwidth requirement and to relay it on a narrower communications channel, as will be discussed in the teleconferencing chapter.

Quantization Error

The digitization process may introduce another problem, a quantization error, if not enough levels represent the analog signal. If, for instance, a video system is governed by a code based upon a two-bit word, only four distinct colors would be reproduced. The original analog signal and the scene the camera is shooting would not be accurately represented.

To correct the problem, the number of levels can be raised, but this option increases the signal's bandwidth requirement. If relayed, the signal may have to be placed on a wider communications channel to accommodate the additional information. Thus, a compromise is usually made between the accuracy of the digitization process and the signal's channel requirement.

Dominance of the Analog World and Standards

Another disadvantage of digital technology is that we live, to a certain extent, in an analog world. Many forms of information, in addition to devices and systems that produce and relay information, are analog in nature. These include telephones, televisions, and radios. This state of affairs necessitates the use of ADCs and DACs.

The Public-Investment Factor

The public-investment factor must also be weighed. If the communications system is switched to an all-digital standard without a transitional period, our telephones, television sets, and radios would have to be replaced with digital-compatible units or used with special converters. The same principle applies to television and radio stations and other organizations that similarly produce and exchange analog information. In sum, this retooling of the communications industry would cost billions of dollars and would disrupt the industry and, inevitably, society.

These are two of the reasons why the change to a digital standard has been evolutionary rather than revolutionary. As described throughout the book, many of the new technologies and their products will be integrated in the current communications structure, which may ease the impact of their introduction.

A case in point is the recording industry and the compact disc (CD) market. If an individual owned a standard turntable and a large record collection, an overnight conversion to a CD-based format would have made the original system obsolete. The old records, the LPs, were not compatible with a CD player, and the owner may not have been able to purchase additional records for the original system.

Accordingly, CD players and discs were integrated in the record industry. Companies began manufacturing a range of CD players while still maintaining their conventional equipment lines. The record industry likewise supported the new medium but continued to produce standard vinyl LPs.

As more CD players were sold, the market for CDs expanded. Record companies increased their CD production and curtailed the number and variety of standard vinyl albums. This trend accelerated, and the digital format, along with audiocassettes, dominated the industry. But, since the process was somewhat gradual, the industry wasn't disrupted. Owners did not suffer immediate losses, and equipment manufacturers continued their support of LP systems, albeit at a reduced level.

Other factors do play a role in the acceptance of a new technology. This includes whether or not a product can be manufactured cost-effectively in comparison with the product it is replacing.

For the CD, the cost for a player sharply fell over a relatively short time period. This prompted more people to buy the systems, which spurred the manufacturing of more discs. As additional discs became available, more people became interested in CD technology and bought players. This, in turn, further motivated record companies to increase their CD lines.

Another potential factor in the acceptance of a new technology is whether or not its products are superior to preexisting ones. The CD and CD players satisfied this criterion, which contributed to their popularity.

Finally, while the gradual integration of a

new technology and its products may benefit the consumer and certain industry elements, it may not always be the best technological solution. The issue of backward compatibility, for instance, somewhat chained the current color television standard to the past, to the detriment of a picture's perceived quality. Similar concerns were raised that this same scenario could have been played out in the development of a new television standard, as discussed in Chapter 9.

On the one hand, a bold, technological step forward, free of past constraints, could revolutionize elements of the communications industry. The downside, though, could be industry disruption and the public-investment factor.

STANDARDS

This final section is devoted to the idea of standards in the communications field, another central concept that will resurface in future chapters. Simply stated, standards are a series of technical parameters that govern the operation of various communications equipment and systems. The standards dictate how information is generated, stored, and exchanged.

A number of domestic and international organizations are engaged in the task of establishing standards. These include the Society of Motion Picture and Television Engineers (SMPTE), the International Organization for Standards (ISO), the FCC, and the International Telegraph and Telephone Consultative Committee (CCITT).

The influence of a standard upon a communications system can vary. Some standards are mandatory and legally enforced. Other standards are voluntarily supported to avoid the chaos that would arise if multiple standards were adopted by different organizations.

Standards may also be de facto in form, as is the case with stereo television in the United States. Even though the FCC did not select a specific standard for this service, a single standard emerged for various economic and technical reasons, and it was supported by the industry.

The idea of allowing an industry to adopt its own standards, especially in the telecommunications industry, is becoming more prevalent. In the view of the NTIA,

The task of standards-setting is best left to the private sector. . . . We recognize that there may be rare cases where FCC or NTIA action to expedite the standards process could be justified. . . . Government intervention could include mediation among conflicting interests or a mandate to industry to develop standards by a time certain [sic], leaving the actual development of standards to the private sector. . . . Any intervention, however, should be limited to cases where there is a specific and clearly identified market failure, and where the consequences of that failure outweigh the risk of regulatory failure (i.e., forcing the resolution of a standard too early in the development of a technology).[22]

Thus, the private sector, industry, should have the freedom to develop the appropriate standards in an open market. However, if standards do not emerge, to the detriment of a technology and an industry, then government intervention may be appropriate.

The adoption of this regulatory stance would be especially important for new technologies and their products that had to compete in a marketplace with established competitors. As briefly described in the next section and in more detail in Chapter 10, this model of government regulation and intervention could have potentially helped the developing teletext industry in the United States.

Importance of Standards

The standards issue is vital for several reasons. First, if standards did not exist, it would be almost impossible to develop any type of electronic communications system and to foster the idea of equipment compatibility. Without standards, the telephone and television you buy may not be compatible with the local telephone system and the signals broadcast by the local television station.

Second, standards can promote the growth of the communications system. If a standard is adopted, both manufacturers and consumers can benefit from this decision. A manufacturer would be assured that its television set would work equally well in New York City and Alaska. The consumer would be more willing to purchase the set for the same reason.

There can also be two or more standards in the same industry, and depending upon the circumstances, they may not adversely affect the industry and hinder its overall growth. An example of this situation was the parallel development of the Beta and VHS videotape formats. While incompatible, they nonetheless created their own market niches. This holds true even though the Beta standard diminished in its market performance, and VHS became the evolutionary de facto consumer standard over time.

The presence of two strong standards in a market may also accelerate an industry's overall development. The competition may spur a manufacturer to introduce improved equipment to gain a larger market share.

Yet, while multiple standards may sometimes be beneficial, a desirable outcome is not always fulfilled. As will be explored in later chapters, this factor could have a devastating effect upon organizations that are attempting to integrate a new and developing technology in the communications field. For teletext services, the presence of more than one standard splintered the market and hindered the potential growth of teletext systems in the United States. A similar scenario hindered the development and growth of ISDN systems.

A third reason for the importance of standards is consumer and industry protection. An accepted standard helps guarantee that a piece of equipment you buy today will not be replaced by a new and incompatible device tomorrow. While new developments do take place and equipment may become obsolete, this process may be a gradual one, as is the case with CDs.

It is also appropriate to point out that obsolescence does not necessarily mean incompatibility. An older videocamera may not incorporate all the latest features of the current generation of cameras. But it may still be quite useful and compatible with your system, even though the pictures may not be as sharp as those produced by later models.

There is a point, though, where a standard may be completely replaced by another standard. An example of such a situation was the abandonment of the 1/2-inch black-and-white video tape recorder (VTR). In the 1970s, this portable VTR and its companion camera were very popular with video artists and schools. The camera was fairly simple, in terms of its design and capabilities, and the recording deck used reel-to-reel tapes instead of cassettes.

Despite the system's popularity at the time, it became obsolete. The equipment and the standard by which it was governed were superseded by other standards that supported new families of hardware and videotape formats.

Another important point about standards, in regard to the international video and television markets, is the number of incompatible standards. The industry in the United States operates under the system devised by the National Television Standards Committee (NTSC), also called the NTSC standard. The standard dictates, among other properties, the number of pictures produced by a videocamera each second, the number of lines of resolution in a picture, and how the picture information is coded in signal form.

Outside of the United States and those countries that have adopted the NTSC system, other standards are employed. Two of these are the French, Sequence a Memoire (SECAM), and the English, Phase Alternation Line (PAL), systems, each adopted by a number of countries in addition to their nation of origin. All three standards are incompatible, and a video program produced under one system must undergo a conversion process to make it compatible with another system.

This arrangement poses some problems for the international market. It creates a roadblock in the exchange of programs between countries due to the mandatory conversion step. It also adds to a program's overall cost.

Thus, even though the adoption of a single standard might have fueled the growth of a country's domestic video industry, the variety of standards are a hindrance on the international level for current systems and for the introduction of new technologies. This problem has also assumed a degree of greater importance as the communications industry becomes increasingly international in scope. This is in keeping with the automotive, computer, and other major industries that have turned toward the international market for sales, manufacturing, and investment opportunities.

REFERENCES/NOTES

1. Ken Marsh, *The Way the New Technology Works* (New York: Simon and Schuster, 1982), 26.
2. Frederick Williams, *The New Communications* (Belmont, Calif.: Wadsworth Publishing Co., 1984), 133.
3. Simon Haykin, *Communication Systems* (New York: John Wiley & Sons, 1983), 6.
4. Tom Smith, *Telecabulary 2* (Geneva, Ill.: abc TeleTraining, Inc., 1987), 28.
5. "Al Sikes's Grand Agenda," *The New York Times*, June 2, 1991, Sect. 3, p. 6. Please note: All over-the-air communications systems do not require specific FCC allocations. Commercial over-the-air lasers discussed in Chapter 11 fall in this category.
6. Basically, the public would still be served by various types of programming, such as news and public-information shows.
7. Ayn Rand, "The Property Status of Airwaves," The Objectivist Newsletter 3 (April 1964): 13; from the reprint The Objectivist Newsletter, vols. 1, 2, 3, 4 (New York: The Objectivist, Inc., 1971).
8. Gary M. Kaye, "Fiber Could Be Winner in the Battle for the Spectrum," *Photonics Spectra* 25 (November 1991): 79.
9. Technically speaking, the bandwidth is the difference between the highest and lowest frequencies, the range of frequencies a communications channel can accommodate. The term *bandwidth* is likewise applied to the signals that are relayed over these channels. It should also be noted that a time factor plays a role in this system. In essence, a communications channel can accommodate or relay a specific volume of information in a given time period based upon the channel's capacity and the noise present on the line.
10. Bernhard E. Keiser and Eugene Strange, *Digital Telephony and Network Integration* (New York: Van Nostrand Reinhold Company Inc., 1985), 19.
11. William Flanagan, "Digital Voice and Multiplexing," *Communications News*, March 1984, 38E.
12. Thanks to Jim Loomis, director of Telecommunications Facilities, Ithaca College, for his suggestions for this section.

In respect to a more detailed look at this process, three phases are initiated and completed. In the sampling phase, the analog signal (continuous in time and amplitude) is sampled. This creates a Pulse Amplitude Modulation (PAM) signal. The amplitudes of the individual and discrete PAM pulses produced at this time correspond to the variable amplitude of the original analog signal at the sampling points. Thus, the amplitudes of the PAM pulses are continuously variable, like the original signal. In the next phase, the quantization stage, the PAM pulses are assigned to the nearest steps or levels in reflection of their amplitudes. This phase converts the wide range of amplitudes to a finite and limited number of amplitudes or values. The final phase is the coding process. At this time, the samples are coded in binary form. Consequently, the original analog signal, which is continuous in time and amplitude, is made noncontinuous by the sampling and quantization phases.

In this type of system, the information is also composed of pulses with identical amplitudes. This is a reflection of PCM information where the amplitudes do not convey the information.

13. The T1 line is only one of a digital family of lines.
14. Arch C. Luther, *Digital Video in the PC Environment* (New York: McGraw-Hill, 1989), 51. The picture is organized in memory as a series of pixels. Please note: See Frederick J. Kolb, Jr., et al., "Annotated Glossary of Essential Terms," *SMPTE Journal* 100 (February 1991): 122, for specific technical details.
15. Tom Smith, *Anatomy of Telecommunications* (Geneva, Ill.: abc TeleTraining, Inc., 1987), 107.
16. Michael Mirabito, *The Exploration of Outer Space with Cameras* (Jefferson, N.C.: McFarland and Co., 1983), 30. Please note: Digital communications systems have similarly played an important role in later missions. As described in Chapter 9, digital technology also provided us with the tools to enhance and manipulate the pictures produced by a spacecraft's camera system after the data were received on Earth.
17. The terms *bits per second* and *baud* are associated with a modem's relay speed. There are differences between the two, however. Please see Brett Glass, "Buyer's Advisory," *InfoWorld* 13 (October 21, 1991): 147, for a more detailed look.

18. Don Wiley, "The Wonders of ISDN Begin to Turn into Some Real-World Benefits as Users Come On Line," *Communications News,* January 1987, 29.
19. Bill Baldwin, "Integrating ISDN Lines for Financial Users," *Telecommunications* 25 (June 1991): 34.
20. NTIA, "Telecommunications in the Age of Information," NTIA Special Publication 91-26, October 1991, 184.
21. "Dropping the Color LCD Tariff Will Save Jobs," *PC Week* 9 (February 10, 1992): 76. Please note: American chip manufacturers experienced a recovery during the early 1990s due to various circumstances (for example, a consortium of companies to promote research and development).
22. NTIA, "Telecommunications in the Age of Information," NTIA Special Publication 91-26, October 1991, xvi.

ADDITIONAL READINGS

Bigelow, Stephen J. *Understanding Telephone Electronics.* Carmel, Ind.: SAMS, 1991. A comprehensive overview of the telephone, from basic concepts to digital and network operations.

Boult, Raymond. "Digitization of Public Net Puts France in ISDN Lead." And, Valigra, Lori. "Japan ISDN Services Attract Leading Users." *Network World* 7 (April 30, 1990): 31, 34. Both articles cover ISDN developments in France and Japan, respectively.

Clement, Fran. "Digital Made Simple." *Instructional Innovator,* March 1982, 18–20. A good primer about digital information and signals.

Edwards, Morris. "What Telecomm Managers Need to Know About Europe '92." *Communications News,* June 1991, 64–68. An examination of the state of the European communications system.

Felker, Lex. "FCC Works Spectrum Shift." *TV Technology* 9 (May 1991): 12. And, Messmer, Ellen. "NTIA Serves Up Plan for Auctioning Radio Spectrum." *Network World* 8 (March 4, 1991): 4, 54. An examination of various spectrum issues.

Haykin, Simon. *Communication Systems.* New York: John Wiley and Sons, 1983, 408–428. Technical descriptions of PCM and multiplexing operations, among other communications topics.

Kishan, Shenoi. "The Many Flavors of T-1 Compatibility." *Network World* 4 (May 18, 1987): 35–36, 46. Examines the T1 carrier in relation to network compatibility.

Posa, John G. "Phone Net Going Digital." *High Technology* 3 (May 1983): 37–43. A comprehensive look at the telephone system and its conversion to a digital format. Various services are also examined.

"Radio Wave Propagation." *Hands-On Electronics,* November/December 1985, 32–38, 104. A comprehensive primer on how a radio wave can travel from its source to the final destination, a receiver.

Reynolds, George W., and Donald Riecks. *Introduction to Business Telecommunications.* Columbus, Ohio: Merrill Publishing Co., 1988. And, Thomas, Ronald R. *Understanding Telecommunications.* Blue Ridge Summit, Pa.: TAB Books, Inc., 1989. Both books provide a broad overview of the telecommunications industry.

Rogers, Tom. *Understanding PCM.* Geneva, Ill.: abc TeleTraining, Inc., 1982. One of a series of books and booklets published by abc TeleTraining. This particular publication provides an excellent overview and an explanation of PCM. It also describes the complete conversion process, from analog to digital, and back again, in terms of all of the steps in this process.

Stallings, William. "CCITT Standards Foreshadow Broadband ISDN." *Telecommunications* 24 (March 1990): 29–41. An examination of B-ISDN, its potential applications, and technical considerations.

Streeter, Richard. "Is Standardization Obsolete?" *Broadcasting,* February 9, 1987, 30. A brief about the importance of standards.

Thomas, Mike W. "ISDN: Some Current Standard Difficulties." *Telecommunications* 25 (June 1991): 40–48. An examination of ISDN progress and standards implications.

GLOSSARY

Analog signal: A continuously variable and varying signal. The communications devices and systems we are the most familiar with, such as videocameras and radio stations, both produce and process analog signals.

Analog-to-digital converter (ADC): An ADC converts analog information into a digital form. ADCs work with DACs to bridge the gap between analog and digital equipment and systems.

Binary digit (bit): A bit is the smallest piece of information in a digital system and has a value of either "0" or "1." Bits are also combined in our communications systems to create codes to represent specific information values.

Channel: A communications line; the path or route by which information is relayed.

Communications channel's bandwidth: A communications channel's bandwidth, its capacity, dictates the range of frequencies, and to all intents and purposes, the categories and volume of information the channel can accommodate in a given time period.

Digital signal: Unlike its analog counterpart, a digital signal is not a continuously variable and varying signal. It assumes only a finite number of discrete values, and digital information is represented by bits, 1s and 0s.

Digital-to-analog converter (DAC): A DAC converts digital information into an analog form.

Electromagnetic spectrum: The entire collection of frequencies of electromagnetic radiation.

Frequency: The number of waves that pass a given point in a second. The frequency of a signal is expressed in cycles per second, and more commonly, in Hertz.

Integrated Services Digital Network (ISDN): A digital communications platform that could seamlessly handle different types of information (for example, computer data and voice).

Modem: The device that makes it possible to relay computer information over a voice-grade telephone line. A modem is used at each end of the relay.

Modulation: The process by which information is impressed upon a carrier signal for relay purposes.

Multiplexing: Multiplexing is the process whereby multiple signals are accommodated on a single communications line.

Picture element (pixel): A pixel is a segment of a scan line.

Pulse Code Modulation (PCM): A digital coding system.

Repeater: A repeater strengthens or boosts a signal in the course of a relay.

Standards: The technical parameters that govern the operation of a piece of equipment or an entire industry. A standard may be mandated by law, voluntarily supported, or de facto in nature.

T1 line: One of the workhorse and important digital communications channels. Information is relayed at a rate of 1.544 megabits/second.

Transducer: A transducer changes one form of energy into another form of energy. Transducers are the core of our entire communications system and include equipment such as microphones and videocameras.

3 Computer Technology Primer

The computer has played a vital role in the launching of the communication revolution. Specifically, the development of the microcomputer, also known as the *personal computer* (PC), has enabled individuals and organizations to use this tool for jobs ranging from word processing to creating spreadsheets that can chart a company's profit-and-loss margin. Other applications include designing expert medical systems and creating graphics for a nightly news program.

Besides such operations, computers and microprocessors have contributed to the creation of a new generation of television and radio equipment. A *microprocessor* is a "computer central processing unit (CPU) built as a single tiny semiconductor chip. . . . It contains the arithmetic and control logic circuitry necessary to perform the operations of a computer program."[1] For our purposes, a microprocessor is a small yet powerful integrated circuit that can be built into audio and video equipment to help perform a series of functions. A PC, for its part, contains a CPU and internal memory. The PC can also input (receive) and output (relay) information.

As stated, both microprocessors and computers have been interfaced with audio and video equipment to speed up various operations and to help human operators in their work. A microprocessor can assist an engineer in "tuning up" a videocamera so it produces a technically superior picture, while a computer can be programmed to automate a series of tasks.

In sum, this chapter is an introduction to computer technology and provides a foundation for the exploration of the computer's role in our communications system. But be-fore we begin, four points concerning terminology and other matters, in respect to this book, must be made.

1. The *generic* term *personal computer* or PC includes a range of microcomputers generally designed for individual use. IBM PCs and IBM-PC compatibles or clones, Commodore's Amiga, and Apple's Macintosh series are all PCs. Due to their widespread integration in the general communications field, this book focuses on these particular computer platforms. By extension, the focus is also on PCs.[2]

2. An IBM clone refers to third-party computers that conform to the IBM microcomputer line's technical specifications. These computers can tap the enormous IBM software universe and can use many of the same hardware components. For our discussion, the *generic* term *IBM PC* is used when describing this family of computers as well as actual IBM PCs.

3. The terms *data* and *information* are used as defined by Alan Freedman in his valuable reference work *The Computer Glossary*. Information is "the summarization of data. Technically, data are raw facts and figures that are processed into information. . . . But since information can also be raw data for the next job or person, the two terms cannot be precisely defined. Both terms are used synonymously and interchangeably."[3] This book follows suit.

4. When appropriate, specific hardware components and software packages are used to describe various computer applications. The products have been selected as representative examples of how, for example, a

Figure 3–1
The Quadra 950, one of the Macintosh line of personal computers. (Courtesy of Apple Computer, Inc.) [photo credit: John Greenleigh]

graphics program can support a given application. Consequently, the hardware and software serve as springboards to explore broad application areas rather than solely focusing on individual products.

COMPUTER HARDWARE

The PCs used in the communications industry are equipped with a number of the same components. Some of the important ones, forms of which may be used by other computer systems, are as follows.

Memory

A computer's memory is its internal storage system and is considered to be a work space. The most important type of memory is the *random-access memory* (RAM), and it is typically expressed in kilobytes, represented by the letter *K*. A computer equipped with 48,000 bytes (48K) of memory can, roughly speaking, hold or store 48,000 letters, numbers, and other characters.

PCs suitable for many of the applications described in this chapter are usually furnished, at a minimum, with 640K of RAM. Many computers exceed this limit and may have one or more megabytes (MBs) of memory.

As indicated, the RAM is the computer's

work space. When a program is loaded in the computer, it is stored in the RAM. When data are subsequently manipulated, such as a letter written with a word-processing program, the letter is likewise stored in RAM. Thus, a computer must be equipped with a specific quantity of RAM to run a given program, and most programs perform more effectively if they can address or use memory beyond the minimal RAM requirement.

The RAM work space is also considered a temporary storage area. Once the computer is turned off, the data in the RAM are cleared from the memory and cannot be recalled unless the data were previously saved on a permanent storage system.

Central Processing Unit

The CPU is the computer's brain. It also dictates the type of software the computer can run, and it plays an important role in determining how fast the computer processes data. Other contributing factors that affect a PC's overall performance include the CPU's memory-addressing capabilities, the speed of the PC's storage systems (such as a hard drive, discussed in a later section of this chapter), how efficiently a program is written, and the use of special auxiliary chips, including math coprocessors.

A math coprocessor can speed up certain operations. In the case of a graphics program written to take advantage of the chip's capabilities, an image may be completed in a shorter time period.

Expansion Boards

Some computers are designed with internal slots that can be fitted with expansion boards or cards. The boards either support basic functions, such as generating a display for a monitor, or enhance the computer's operation. They may be required or may supplement and complement built-in capabilities.

A video display board for an IBM PC, for instance, can dictate the type of monitor that is used and various technical parameters. These include the resolution of the graphics and *alphanumeric* information as well as the number of colors that may be simultaneously viewed. Alphanumeric information encompasses alphabetic and numeric characters, that is, letters and numbers.

Other types of boards include modem and memory boards. As implied by their names, modem and memory boards serve as internal modems and memory-expansion products, respectively. A PC can also be fitted with a fax board or even a fax/modem combination.

A *facsimile* or fax machine can receive and send electronic documents over standard telephone lines. Fax technology helps speed up business and other forms of communication. Instead of sending a letter to a customer through the mail, you can fax the letter in a matter of seconds. The customer subsequently receives the letter on his or her own fax machine.

You can buy a dedicated, stand-alone unit or a PC fax board, which may operate in a background mode. When you are working on another program, the board and controlling software automatically and transparently receive any incoming documents. You can also compose a message with the PC, preview the fax on your monitor before sending it, and use your printer to generate hard copies.

Finally, it should be noted that not all PCs are fitted with expansion slots, and some may have only limited internal expansion capabilities. This may be a reflection of design, cost, and/or size constraints.

Despite what appears to be a restricting factor, powerful PCs with such design characteristics have been created. A computer may already perform many of the operations that would mandate the use of multiple boards with other PCs. There may also be some expansion capability through external ports or connections.

In the case of the Amiga 500 PC, an external hard drive could be connected to the computer. A hard drive is a data-storage device, and in the case of the Amiga 500, the housing may have additionally been fitted as a memory expansion peripheral.

Besides external capabilities, third-party manufacturers have an uncanny ability to design around internal expansion limitations. Manufacturers have used different techniques to expand and enhance the Amiga 500's processing speed, internal memory ca-

pacity, and other characteristics. But these operations entail opening up the computer case and may void the manufacturer's warranty. PCs designed for user access to the expansion slots generally don't operate under the same warranty guidelines.

Data Storage

As indicated, the RAM is a temporary storage area. Consequently, a system must be used to safely store a program's data before the computer is turned off. The system must also be capable of reloading the data when necessary.

There are a variety of data-storage configurations, including those based on optical-disc technology (as described in Chapter 4). But as of this writing, the most popular and common systems are floppy and hard disks, both of which are magnetic in nature and are governed by magnetic principles.

Floppy disks are used to store data and to distribute programs. There are two primary sizes, 3 1/2 inch and 5 1/4 inch, and a disk's storage capacity, like a computer's memory, is measured in either kilobytes or megabytes. Typical disk capacities are 360KB, 720KB, 1.2MB, and 1.44MB. The device that actually stores and retrieves the data from the floppy is the disk drive.

The second type of storage medium is the hard drive. It holds a large volume of information, equivalent to 100 or more floppy disks, and generally has a capacity of 40, 80, or 100 or more megabytes. A hard drive is faster than a floppy drive in accessing and storing data, and this speeds up an average work session. The drive is usually mounted in a computer case, or in some instances, it can be an external unit.

Monitors

A monitor is the computer's display component. There are several general categories of monitors, and it's important to match a monitor with the appropriate video display board. This combination, in turn, should be matched with the PC and the projected applications.

Monochrome monitors, for example, are

fairly inexpensive, well suited for word processing, and may feature an amber-on-black display. A color monitor, on the other hand, is a superior graphics display device and can additionally support text-based applications.

There are a number of display standards in the PC world, including the VGA (640 × 480 resolution) and Super VGA (800 × 600+ resolution) standards for IBM PCs. There are also special board/monitor pairs for specific applications, including the large and high-resolution monitors that are suitable for desktop-publishing operations (as discussed in Chapter 7).

One of the most dramatic developments in the overall PC market is the advancement of display technologies. The resolution, or for our purposes, an on-screen image's apparent sharpness, is increasing, and contemporary display boards and monitors are easier on the eyes. A monitor may produce a higher-resolution display with less flicker, to help reduce eyestrain.

A video display board may also be designed to optimize and speed up the performance of different software packages, such as Microsoft Corporation's Windows, described in a later section of this chapter. Video boards are also fitted with their own memory, and as the memory increases, so too does the number of possible colors that can be simultaneously displayed at different and higher resolutions.

At this time, 24-bit systems have become popular. The 24-bit figure refers to "a display standard in which the red, green and blue dots that compose a pixel each carry 8 bits of information, allowing each pixel to represent one of 16.7 million colors."[4] Basically, instead of working with a limited palette or range of colors, 24-bit systems can gain access to a palette of millions of colors. The actual number of on-screen colors is lower, though, than 16.7 million.

Visually, a picture is more vivid and aesthetically pleasing when displayed on such monitors. When used with appropriate software, an image, such as a landscape with clouds, may appear to be lifelike or photorealistic.

THE MICROCOMPUTER

Prior to the PC era, the computer industry was dominated by two types of systems, the mainframe and the minicomputer. Both systems have been and may still be used by organizations with multiple users and extensive processing needs. But there is a price for this sophistication and processing power, especially in regard to mainframe units: they are much more expensive than PCs.

The popularity of the PC is due, to a great extent, to the introduction of the original, widely available, IBM microcomputer in 1981. People were already familiar with IBM products, and the corporation's entry in the microcomputer field legitimized the PC in the eyes of the business community. More advanced models were later created by IBM and other manufacturers, and powerful software packages, which played a complementary and major role in the launching of the PC revolution, were developed.

The PC's low cost, in comparison with minicomputers and mainframe units, also made computer processing accessible to a broader user base. Individuals purchased PCs to increase their productivity, as did small and medium-size businesses that could not afford to computerize their operations up until this time.

As the user base expanded, the price for PCs fell as more models were mass-produced. PCs with enhanced processing capabilities and the accompanying software that tapped this computational power also attracted new, as well as established, computer owners. All these factors, including the cost-effectiveness inherent in digital components, contributed to the growth of the computer industry, and inevitably lowered a PC's price. In fact, by the 1990s, it was possible to purchase a computer that was far more sophisticated, but less expensive, than earlier models. Two computer ads illustrate this point.

In a December 1982 issue of *Byte*, a popular computer magazine, an ad from a mail-order company listed the price for an Atari 800 computer, equipped with 16K of RAM, for $689.95.[5] This price did not include the separate $469.95 disk drive or any other major components.

In 1992 a merchandise discounting company, which sold products ranging from beds to computers, listed an IBM-PC compatible for $849.99.[6] This price included 1MB of memory, floppy and hard disk drives, and a color monitor.

So, in the course of 10 years, even nonspecialized discount companies sold significantly more-advanced hardware for less money. The computer industry is one of the few manufacturing areas where this type of price vs. performance ratio prevails, to the users' benefit.

The latest PCs are also the ideal tools for many of the applications common to the communications field. There is an extensive base of general-release and specialized software, such as word-processing and radio-station-management programs, and these applications may all be supported by the same computer. This is one of the reasons why PCs have forever altered the way we communicate, work, and play.

Portable PCs

In addition to desktop PCs, manufacturers produce portable, battery-operated *laptop computers*. A laptop can be used in the field by a reporter covering a story. After the story is written, it can be relayed to the newspaper via a modem and a telephone line. An operation of this type enhances the communication process, since information can be rapidly transmitted from one site to another. This same scenario also applies to the business world.

Laptop computers are equipped with a variety of display systems. You typically open the top of a case to reveal and use a special screen, which may support a gray-level or color display.

A laptop is also usually fitted with floppy and hard drives as well as an internal modem. Other features may include an external monitor port, the capability to expand the computer's memory, and a power-saving option to help conserve the internal battery.

Figure 3–2
A typical PC application: producing graphics for video productions as well as for presentations. (Courtesy of Impulse Inc.; created with Impulse's Turbo Silver program)

Laptops have spawned even smaller portable PCs called *notebook computers*. A notebook PC generally weighs several or more pounds compared to 10-plus pounds for many laptops, and are ideal for situations where a laptop may be too large or heavy. A number of notebooks and, if applicable, laptops can also be fitted with desktop docking stations, providing additional expansion capabilities.

In keeping with this downward or miniaturization trend, *palmtop computers* are becoming more popular. As implied by the name, palmtops are handheld computers. They may run standard or customized versions of commercial software packages and typically use special memory cards for data storage.

Palmtops fill a unique niche in the PC world. Their keyboards and screens are small, and they have limited processing power. But they can be used and carried in places where even notebook PCs may be too cumbersome. Palmtops can also operate for extended periods without a change of batteries, and there are provisions for exchanging data with other PCs.

While this miniaturization trend will continue, there are, at least for the near future, size constraints in designing portable computers. These constraints may have more to do with the human–PC interface than with the capability to design ever-smaller computers. In general, as keyboard and screen sizes decrease, their effectiveness, at least for the operator, may decrease in kind. A partial solution may lie in adopting more advanced interfaces, such as pen-computing and speech-recognition systems, as discussed in later sections of this chapter.

Human–Computer Interface

One important consideration in using any computer is the human–computer interface: how we communicate with and control the computer. For this discussion, interfaces are the hardware tools we use to work with computers. They range from keyboards to graphics tablets to virtual-reality systems. (The latter interface, where we can interact with a computer and actually become part of a computer-generated world, is examined in the book's afterword.)

Software is the other piece, the driving force behind human–computer interfaces. A computer's operating system provides the most basic link. Other software interfaces, especially those that are interactive in nature, are covered in Chapter 4.

As indicated, there are a variety of interfaces on the market, the most common of which is the *keyboard*. Selecting a keyboard is a personal matter. Keyboards feel different depending on the design, and certain keys may be located in different locations. For IBM PCs, function keys, which are used by word-processing and other programs for various operations, can be placed on the top, side, or in both keyboard locations. The placement can help speed up typing for those used to working with a specific keyboard configuration.

A second category of interfaces supplements and complements the keyboard. These include the *mouse, graphics tablet, trackball, light-pen, touch-screen*, and a new PC category, *pen-based computing*.

A mouse is a small, rectangular device that is interfaced with the computer and sits on a desk. As you move the mouse, a cursor on the monitor's screen moves correspondingly. In one application, you can use the mouse to select a command listed on a drop-down menu. As the mouse is moved to the top of the screen, a menu with a list of commands may appear. You highlight a spe-

cific command, click a button on the mouse, and the command is carried out.

In another common operation, a mouse functions as a drawing tool for graphics software. Instead of a keyboard, a mouse serves as the input device. To create a square, you simply trace the shape on the desk. With a keyboard, you would have to press the cursor control keys a number of times to complete the figure.

A more effective tool for this task is the graphics tablet, a drawing pad interfaced with either a puck or a pen. You can literally sketch a picture in freehand, among other options, with this device. Some configurations also provide very fine degrees of control when combined with specific software.

The graphics tablet is a more natural drawing interface than the mouse. While a mouse may feel, depending upon the model, like a block of wood, a pen is more comfortable. It's closer to a traditional drawing tool in size, shape, and function. To test a graphics tablet's superior handling, try writing your name with a mouse and then with a pen and tablet.

The trackball works much like a mouse. But instead of controlling the cursor by moving the mouse, you move or roll a small ball built into the trackball housing. Because trackballs do not require much desk space, miniature trackballs are popular with owners of portable computers. You can use a built-in or a detachable model in a limited physical area. On the downside, trackballs may be more awkward than mice for certain software operations.

The last two interfaces are light-pens and touch-screens. A light-pen is a handheld device typically used to select items displayed on a monitor's screen. Unlike a mouse or graphics tablet, a light-pen lets you directly interact with the monitor.

The touch-screen enhances this process, eliminating the use of a manual tool to interact with the computer. In a standard application, a menu with a list of commands appears on the screen. You select the command by touching the screen at the appropriate location.

In another application, an interactive

computer program for a museum can be created. When a visitor touches one of the museum's rooms displayed on a monitor, the exhibits at each location are listed.

In both cases, the touch-screen is a very intuitive interface, since you don't have to learn how to use it. Simply touch what you want and the hardware and software interfaces automatically complete the task.

The final human–computer interface, a relative newcomer to PC-based computing, uses the pen metaphor: pen-based computing. In one configuration, you hold a computer much as you would hold a clipboard. But instead of writing on paper, you write on a flat panel display with a pen.

This interface can simplify many routine tasks. Industries that depend on paper forms, such as the express mail/package industry, can now use electronic forms and pen-based systems. As deliveries are made, people working in the field can check off boxes on forms and fill in other information lines. The data are then stored and can be readily transferred to a central computer.

Contemporary systems can recognize handwritten text and special markings used for editing. Recognizing conventional cursive or nonprinted text is, however, a more demanding task.[7]

Future pen-based systems may also draw upon the idea of the electronic book. You could carry an electronic manual instead of a paper-based one. The electronic book could be rapidly updated, annotated with your own notes, and integrated with different graphics.[8] Like a touch-screen, this type of interface is very intuitive since we would be using a familiar communication form, basically an electronic version of paper.

Finally, a new generation of what have been called *personal digital assistants* (PDAs) may also support handwriting recognition. Led by Apple Computer and its Newton PDA, this compact pen-based device marries computer technology with the convenience of using a pen and paper. The system also offers a more transparent and natural interface to tap its functions than its typical less sophisticated counterpart, the traditional *electronic organizer*. This small, wallet-size device can

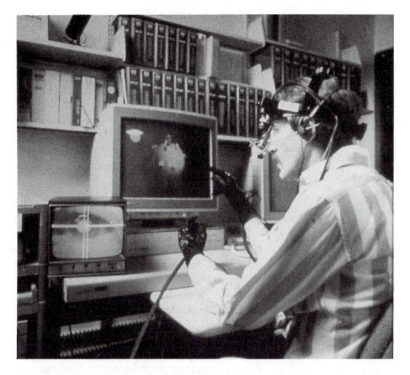

Figure 3–3
Developing advanced and enhanced human–computer interfaces at MIT's Media Lab. (Courtesy of MIT Media Laboratory)[photo credit © Hiroshi Nishikawa]

be used for specialized applications and may be equipped with modules to enhance its capabilities. For example, it can be used as an electronic calendar and, when so equipped, as a foreign-currency-exchange calculator.

COMPUTER SOFTWARE

There are a number of communications software packages available, including general-release programs and those specifically designed for the television and radio industries.

This section briefly describes general software categories. More specific program types, as well as specific programs, are discussed in later chapters in the context of application areas. These include hypermedia and multimedia programs in Chapters 4 and 8, desktop-publishing programs in Chapter 7, and programs designed for the broadcaster in Chapter 9.

Disk Operating Systems and Graphical User Interfaces

The *disk operating system* (DOS) is the most

important piece of computer software. It controls the computer as well as specific data- and file-management and -manipulation operations. Different operating systems have been released for use in a variety of computers.

A standard has emerged in the PC market, however, through the widespread integration of IBM PCs. As of this writing, MS-DOS has dominated the PC world—with a few notable exceptions, including Apple and possibly Amiga PCs and their respective software and operating environments.[9] MS-DOS supports a text-based interface: keywords are typed to carry out various PC functions.

In contrast is the *graphical user interface* (GUI), popularized by Apple and its Macintosh line. Apple helped establish what we've come to associate with a GUI:

- a pointing device, typically a mouse
- on-screen menus . . .
- windows that graphically display what the computer is doing
- icons that represent files, directories . . .
- dialog boxes, buttons . . . and other graphical widgets that let you tell the computer what to do and how to do it.[10]

A windowing operation generates multiple windows, one or more framed work spaces on the monitor's screen, that can be moved, resized, or removed by the user. This function can enhance various computer operations, such as using the windows to display the contents of a computer's drives and, potentially, to help simplify file-copying procedures. Depending on the software, files can be copied without typing a series of keywords. Windows can also be used to switch between different programs displayed in their own work spaces on the monitor.

An icon is a small picture, a pictorial representation of, for example, a disk drive or a file. Instead of typing a command to display the files on a drive, you move an on-screen cursor to the appropriate icon and click a mouse button. Instead of typing a command to delete a file, you move the appropriate icon to another icon, which in this case may be a trash can. The file is then deleted.

Apple's popularization of the GUI has carried over to other PC platforms. There have been a variety of GUIs on the market with different capabilities, including the X-Window System and, specifically for IBM PCs, Presentation Manager and Windows.[11]

In fact, as described at the end of this chapter, Apple Computer has been involved in a multiyear lawsuit with the Microsoft Corporation over an intellectual-property issue. Apple has contended that Microsoft borrowed the *look and feel* of its GUI when Microsoft developed Windows.

Database and Spreadsheet Programs

A database program organizes information ranging from a consultant's client list to the titles of a radio station's carts. Once the information is entered in the computer, it can be organized and manipulated. Song titles, for instance, can be filed and stored under specific categories, including a singer's or a group's name, or even the types of music a station plays.

More sophisticated database programs support a function that goes beyond the simple compilation of information. A program can be used to define and examine the associations between different categories of information. For the radio station, these can include the names of the station's sales staff, their clients and types of businesses, and the amount of revenue each salesperson generated in a year.

A database may also support graphics and pictures in addition to the standard data forms. This type of program merges a visual

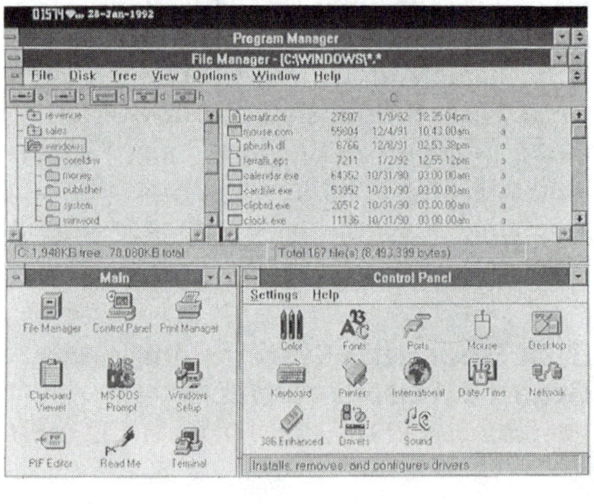

Figure 3–4

Screen shot of Microsoft Corporation's Windows program. Note the icons on bottom half of the image. (Courtesy of Microsoft Corp.)

image with an information file, which may be a written description about the picture. In a typical application, a specification sheet that highlights various features about a house may be integrated with a picture of the site. This configuration would enable a real estate company to maintain a written and pictorial database of houses on the market.

A spreadsheet program, in contrast, is primarily a financial tool. Data are entered via a table format, in columns and rows, and are tabulated and manipulated through a series of built-in mathematical and financial functions.

Working with an electronic spreadsheet, you can rapidly accomplish tasks that would normally require hours to complete. In one application, the spreadsheet serves as a forecasting tool. As various figures on the spreadsheet are changed to reflect higher advertising rates, for example, all the pertinent figures are automatically recalculated. Thus, as you plug in new data, the figures are updated.

A spreadsheet may also support a graphing capability. The data are portrayed pictorially, such as a line graph or a pie chart. Viewing data in this fashion may make it easier to discover relationships between the data. The same graph may also create a more powerful presentation. You can now "see" the numbers instead of just rows and columns on a page.

Word-Processing Programs

A word-processing program is used to write letters, news stories, articles, and other documents. Besides these basic tasks, some programs are equipped with a mail-merge option, merging a mailing or address list with a standard form letter. In this case, if there are 50 addresses on the list, 50 separate letters, each bearing one of the addresses, are produced.

A word-processing program supports other functions that enhance the writing process. You can move designated blocks of text to different sections of a document, and characters, words, and whole pages can be electronically erased. Certain programs can also produce multiple columns of text, import graphics, and may include a thesaurus as well as a spelling and grammar checker.

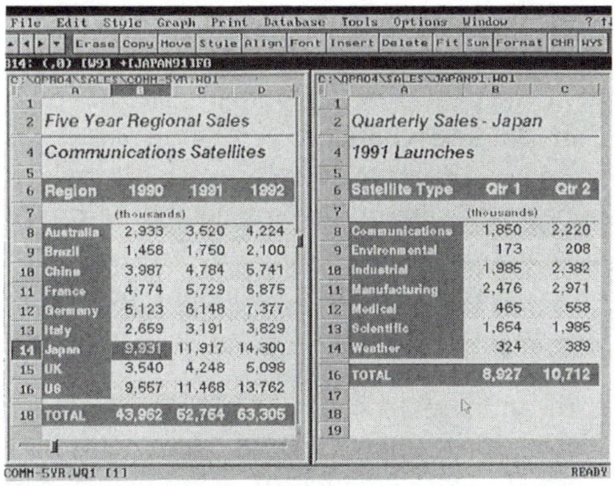

Figure 3–5
Screen shot of a spreadsheet program. Some programs, as in this case, support enhanced design and formatting capabilities. (Courtesy of Borland International; Quattro Pro)

Most word-processing programs can optionally save data in an international standardized code called the *American Standard Code for Information Interchange* (ASCII, pronounced AH-skee). After the information is saved in this format, it can potentially be used by other computer systems and programs, since the ASCII code is universally supported. ASCII is also used in other situations and plays a role in the teletext and videotex systems described later in the book. Alphanumeric information in these systems has essentially been coded in the ASCII standard. Thus, ASCII helps facilitate the exchange of information between computers and, ultimately, their human operators.

Word-processing programs also use control codes, sequences of commands embedded in the text, to trigger special printer functions. Two control codes may, for instance, signal the printer to start and stop its underlining function at specific points in the text. While control codes make it possible for a program to tap a printer's power, the codes are generally not transferable from one pro-

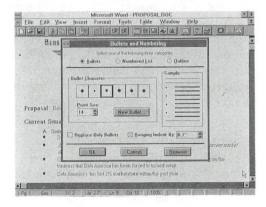

Figure 3–6
Newer word-processing programs support advanced features, including in this screen shot a selection of bullets that could highlight a line of text in a document. (Courtesy of Microsoft Corp.; Word for Windows)

gram to another. The codes are removed or stripped from a text file and must be replaced with the new program's own series of codes if the file is exchanged.

Depending upon the software, though, it may be possible to use a special program to automatically convert one file format to another. A word processor may also be able to support another program's file through an internal conversion process.

In addition to conventional applications, word-processing programs have benefited from the convergence of technologies and applications. Apple, for one, has pioneered the use of voice annotation for word-processing and other software packages.

With a microphone, you can use a PC to record your voice. This message, which may be a comment about a letter, can then be pasted in the document. When someone subsequently reads the letter on a computer, the message can be replayed by clicking on a symbol representing a voice annotation. This particular application marries what is primarily a text-based communication form with voice communication.

Integrated Programs

An integrated program incorporates two or more programs into a single package instead of using separate software for database, word-processing, spreadsheet, and graphics applications. An integrated program also supports data interchangeability. The data from a spreadsheet could be used to create a line graph, or the information from a database could be merged with a word-processing module by a stockbroker writing a letter to a client. Individual packages similarly support this type of information exchange, but several may have to be purchased to duplicate the functions of a single integrated program. The downside to this flexibility, though, is that the various modules in the integrated program may not be as powerful or comprehensive as dedicated software packages.

Programming Languages

The applications programs just described are created through various programming lan-

guages. A programming language is considered a control language in that it provides the computer with a set of instructions. These instructions, in turn, make the computer perform a series of tasks or operations in a specific fashion and order.

Common programming languages include BASIC, Pascal, COBOL, C, Fortran, and Ada. Each language has its own particular strengths and weaknesses and is typically geared toward a select set of applications.

Fortran has a number of functions that makes it well suited for scientific operations, while COBOL is widely used in the business world. BASIC and Pascal are the first programming languages many people learn, and they are designed, in part, with this goal in mind. More esoteric languages, Lisp and Prolog, are the domain of the artificial-intelligence community, and new generations of software that can create multimedia projects, as described in Chapter 8, are appearing on the market.

Ada, a language sponsored by the U.S. Department of Defense, is well suited for working with complex systems. Ada is named after Augusta Ada Byron, considered to be the world's first programmer.[12]

Communications Programs

A communications program enables a computer to exchange information with other computers. When a PC is connected to a modem, the communications software controls specific functions that are central to the relay process. These include how fast data are transmitted and received, as well as other technical parameters.

The programs vary in their capabilities, and it is possible to automate various operations. In a standard application, the system can be set up to dial the telephone number of a computer database company to establish the communications link. A program may also support an emulation mode where a PC can emulate or function as a remote *terminal* so it can be linked with, in most cases, a mainframe computer. A standard terminal is designed to interact with mainframe and minicomputer systems. It consists of a keyboard and monitor, and it is generally less expensive than a PC, since it is not designed to process and manipulate information.

In addition to providing a link between remote computers, a communications program is used when data are exchanged between two computers via a direct hookup. Instead of using a modem and the telephone line, a special adapter, a *null modem,* joins the two PCs, and data can be exchanged at a rate that exceeds a typical telephone connection.

The link for such operations is made through a computer's RS-232 port. This standardized connection defines various technical characteristics that enable a computer to communicate with the outside world.

Graphics Programs

Graphics programs are used to create different types of drawings, such as a company's logo or a rendition of the space shuttle. A graphic can also show us what next year's car model will look like and can chart the population growth in the United States.

There are different categories of graphics programs, some of which are introduced in this section. Beyond the program types and their applications, there are a variety of graphics formats on the market. While these standards may not be compatible, most programs can import or retrieve more than one format, and if necessary, a graphics file-conversion program can be used.

This process also extends, with certain limitations, to graphics created on different PC platforms. The capability to share files ultimately benefits computer users since it facilitates the flow and exchange of information.

Another trend, which may benefit users, is the convergence between program types. A single program may now support multiple functions. Instead of using two or more of the programs listed in this section, a single program may now suffice.

Paint Programs. A paint program addresses or manipulates the individual pixels on a screen. The pixels can be assigned specific colors and can be controlled to produce numerous effects. Applications range from

Figure 3–7
A PC graphics program and one of its drawing options. Some programs may also sport multiple capabilities(Courtesy of Corel Systems Corp.; Corel Draw)

manipulating video-based images to computer art, where an artist paints with an electronic medium rather than one based on paint and canvas.

The paint program creates bitmapped graphics. Bitmapping is, in part, the computer's capability to manipulate the individual pixels that make up a graphic. It also describes the method by which the graphics information is stored. As succinctly stated by Gene Apperson and Rick Doherty, "The bitmapped image is represented by a collection of pixel values stored in some orderly fashion. A fixed number of bits represents each pixel value. The display hardware interprets the bits to determine which color or gray levels to produce on the screen."[13] Thus, the values, which essentially are the pixels' colors, are coded, stored, and eventually retrieved and interpreted by the computer to create the graphic.

A paint program can generate a wide selection of brush shapes and sizes, much like brushes used in traditional art. The brushes are actually used to create an electronic picture that appears on the computer's monitor. The picture may be drawn in a freehand style or tools may be used to create circles and other shapes. Sections of the picture can also be magnified for fine, detailed work, and it may be possible to grab part of the picture as a brush. Once grabbed, the brush can be used like any standard brush for painting. It can also be flipped or resized.

Paint programs usually provide some degree of control over the palette. The number of usable colors varies from system to system, and you can alter the available colors to fit

Figure 3–8
An image-editing program may support a range of special effects. In this screen shot, the two pictures on the right have been manipulated by the software. (Software courtesy of Image-In, Inc.; Image-In Color)

your project. One common method is to select a color and to change its red, green, and blue values by moving RGB slider bars. This alters the color. In the case of many IBM PCs, you may be able to select 256 on-screen colors from a potential range of over 200,000 colors. Other programs as well as platforms can support 24-bit images, as previously described.

Image-Editing Programs. An image-editing program's primary function is that of an image editor: to enhance and edit preexisting gray-scale and color images through software tools. Typical applications are in desktop publishing and video.

This software family can also be viewed as the word processor of images. Just as you edit, move, and copy a body of text, a picture can be cropped, rescaled, rotated, and merged with another image.

Special filters can also be applied to enhance and optimize an image's appearance prior to its printing either on paper or to videotape. In this case, an image, or select

parts, can be sharpened or softened, or other filters can be applied to create additional visual effects. The image could be further manipulated through gray-scale and color editing and correction operations.

A program also includes a paint module, featuring brushes and other tools, and may support a selection of scanners. As described in Chapter 7, a scanner can digitize photographs and other still images, which are subsequently fed to a computer. With this software and hardware combination, a picture can be scanned and then retouched or altered.

Drawing (Illustration) Programs. A drawing program does not address individual pixels. It treats a graphic as a series of individual geometric shapes or objects, such as a circle or a rectangle, that can be manipulated and moved to different screen locations. Graphics are *vector based*, and picture information is expressed and stored mathematically, not as bitmaps.[14]

Drawing programs are used for creating ads, illustrations, and other projects, and have powerful text-handling tools. Words can have a three-dimensional appearance and can follow designated nonlinear paths.

Like paint-type software, drawing programs have their own advantages and application areas in which they excel. For example, when a graphic produced by a drawing program is enlarged or reduced, the lines will remain sharp and well defined.[15]

Computer-Aided Design Programs. A computer-aided design (CAD) program is similar to a drawing program since it manipulates geometric shapes. But it has a number of enhancements that make it a precision tool, extremely powerful for architectural and industrial design.

There are two-dimensional (2-D) and three-dimensional (3-D) CAD programs, the latter of which can be used to create, for example, a three-dimensional view of a building. This image can then be rotated to show the building from different perspectives.

Initially, 3-D views of buildings and other objects were generally limited to wire-frame drawings, images composed only of lines,

Figure 3–9
An example of a house created with a 3-D CAD program. Please note the support for multiple views. (Courtesy of American Small Business Computers; Design-Cad 3D)

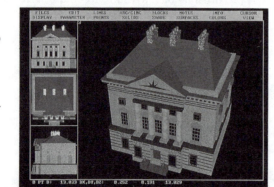

with no solid appearance. Advancements in computer hardware and software have, however, made it possible to break these restrictions, and we can now add physical surfaces and solid attributes to the wire frames. The building in our example would look more like a "real" building from various angles rather than a wire outline.

The same technology has been adopted by the automotive industry; a car can be designed and then viewed from a variety of angles. The vehicle's realistic appearance, in contrast with a wire-frame model, would help the designers assess more accurately its physical characteristics.

A program may also support an animation feature. It may likewise be possible to tie into another program to reveal the area(s) of a designed part that may be subject to stress during an actual operation.

Finally, the information, such as for a car part, could be fed to a series of computer-controlled tools. At this point, an engineer's vision, a drawing on a computer monitor's screen, is transformed into the actual physical part by a computer-controlled tool. This function is an element of computer-aided manufacturing (CAM), an area closely allied to CAD systems. In many instances, CAD and CAM are linked together in a CAD/CAM system.

Animation and 3-D Programs. An animation can be described as a series of images that, when viewed in sequence, convey movement and motion. We're all familiar with Saturday-morning cartoons, and some of the same techniques used to produce Bugs Bunny have been adopted for projects ranging from business presentations to the opening of a video production.

Contemporary programs have simplified the animation process. Some allow you to create predefined paths that an object, such as a figure, will automatically follow. You select the number of frames, determine the starting point, and set up the path or direction(s) of the movement. The program will then draw the individual frames, out of the specified total, as the object progresses down the path.

A program may also support other func-

tions. For example, you may be able to speed up or slow down the animation at specific points, for smoother transitions and enhanced realism. Or you may be able to define two distinct shapes, Shape A and Shape B, and then command the software to draw intermediate frames so that Shape A changes into Shape B during the animation (called *morphing*).

Cycling the colors on a screen is another option. You select a range of colors, and then the on-screen colors cycle through this range. Cycling can create the illusion of motion while only using a single frame.

Animations and still images can also be created in *3-D*, adding realistic depth. When a sphere rotates, it looks like a sphere, not a flattened-out object. A picture can also be wrapped around a three-dimensional object, and textures, such as wood, can be applied.

Programs may support 24-bit images, and you typically have control over light sources and camera parameters. Light sources are individual lights that illuminate a scene. The number of lights a program can support may vary, and it may be possible to scale a light's intensity. The camera parameters affect what we see on the monitor. A view could be changed from wide-angle through telephoto, and in an animation, the camera itself could move.

Lifelike still images and animations can be created through *ray-tracing*, a rendering technique that "literally traces the paths of thousands of individual rays of light through a three-dimensional scene via computation."[16] Ray-tracing can produce very realistic images with accurate shadows and reflections.

The only penalty is time. It may take hours to render a scene, depending upon the PC. But you can initially create the scene in a faster mode. When finished, the scene can be reviewed, changes can be made, and the final sequence can then be produced.

When the work is finished, it can be recorded on a VCR. If the project exceeds the size of your usable memory, you can use a system such as a single-frame controller. Basically, a PC is interfaced with a special VCR, and the individual frames are recorded either from disk or as they're rendered. Various programs support this operation, including popular 3-D animation packages such as Autodesk's 3D Studio.[17]

Presentation Programs.

Presentation programs are designed to generate charts and graphs ranging from a single page to multipage presentations. Templates are available to format the data, or by using a set of tools, you can create your own designs. There are also color, text, and line-style options, among others.

It is possible to import data from other programs. You can, for instance, use the data from a preexisting spreadsheet to create a chart. It may also be possible to link both files. If the spreadsheet is altered, the new figures in the specified range will be reflected in the linked chart. This saves time since the chart is automatically updated.

Presentation programs may also support a screen-show mode. If you are creating a computer-based presentation composed of a number of charts, they can be shown in sequential order. You select the charts, the display order, the viewing time, and, possibly, the transitions between the images, such as wipes or fades. If appropriate, transitions can make the presentation more interesting. When you're finished setting up, the PC runs the show.[18]

Besides this output, presentation programs support standard and color printers. It may also be possible to use an in-house film recorder to produce a high-resolution slide. Failing this, an external service bureau that specializes in this area can create slides from files.

Mapping Programs.

A mapping program generates maps, on-screen graphics of specific geographical regions. The information can include a collection of U.S. street maps and specialized maps that serve as analytical tools when linked with the appropriate information. In the latter application, a map can be used to examine the demographic makeup of a state or locality.

There are also programs designed for consumers. A program can be linked with a database of facts. As you view a map of a country, you may be able to retrieve climatic and other types of information. You may also be able to use a form of mapping program for planning a road trip. Select your starting and end points, and the software plots out the route.

Another type of program may support digital landscape data. You can produce and view accurate renditions of Mt. Saint Helens and other sites, and possibly even other worlds. The final product can be either a still image or an animation.

Visualization Programs

Scientific visualization is the "ability to simulate or model 3-D images of natural phenomena on high performance graphics computers."[19] Data are graphically represented, and this visual representation can provide insights into the functioning of our world, and the universe at large.

Instead of looking at pages of numbers, a scientist can view the data graphically. In one case, while designing a new spaceplane, airflow and thermal characteristics could be readily observed through visualization.[20] Essentially, the visual image makes it easier to interpret the data since you actually see, in a sense, the data brought to life. The invisible is made visible.

Scientific visualization is currently the domain of two classes of specialized, powerful computers, supercomputers and workstations. However, as with most developments in the computer field, there is a slow trickle-down effect from high-end systems to the realm of powerful PCs.

Implications.

To wrap up this discussion about graphics programs, it may be appropriate to talk about creativity and computer systems. There are certain advantages in using computer-based technologies. In the graphics area, a PC with the appropriate software can help you transform an idea or vision into a reality, an actual product. It is the marriage between the conceptual and the concrete.

The same system can help you reach this goal even if you are not a graphic artist. You may have an idea for a poster but may not be adept at creating the 3-D letters your project demands. A computer with the appropriate software may have an option that will create the effect. In this case, the PC is functioning as it should: as a tool. You provide the guiding thought, and the PC helps to implement your idea.

This scenario also applies to other application areas. Chapter 7 discusses desktop publishing. A PC-based system is used to design and print newsletters, brochures, and other documents. You can experiment with different page layouts and mix text and graphics on the same page. As is the case with the 3-D effect, the PC has provided you with another tool to transform an idea into a product.

This concept is an important one. If the PC functions as a tool, then the person using it should at least have a basic grasp of underlying aesthetic principles relative to the job at hand. For example, a PC may enable you to produce an ad. But unless there is a guiding aesthetic principle, the message may be lost in a maze of words and graphics. A PC-based system can also be used in the creation of a video production. But unless you understand the principles associated with lighting, camera shots and angles, audio techniques, and editing, the final product may be an audiovisual mess. You may also own a sophisticated word-processing program that can import graphics and automatically generate front and/or back matter, such as a table of contents. But unless you understand the writing process, your message may be garbled.

A PC may be helpful, but it does not circumvent the writing process. A professional videocamera may produce a technically superior image, but its use alone does not guarantee a successful outcome.

While PCs have made it possible to extend our creative vision, they are only tools. We have to supply the imagination and the skills to use them.

Finally, this use of computers has raised a number of questions. Does a computer dehumanize the creative process or does it enhance it? Should we continue to use traditional pro-

duction methods or should we adopt PC-based systems? Can traditional and computerized methods co-exist?

PRINTERS AND LOCAL AREA NETWORKS

A computer configuration may be composed of various components in addition to the hardware and software described thus far. These include the printer and a network that links two or more computers in a communications system.

Printers

A printer is a device that produces a paper copy (a "hard" copy) of the information stored by the computer. Three types of printers dominate the industry, and there is a fourth category of important, yet less widely adopted, machines.

The dot-matrix printer, the first major category, has been a favorite among users for a number of years. It produces alphanumerics and graphics through a matrix of closely spaced dots. The dots are created by a series of pins or wires that strike against a ribbon and onto the paper.

There is a relationship between the number of pins in a print head, the component that houses the pins, and the quality of a printer's output. A dot-matrix printer that incorporates a 24-pin print head, for instance, produces a print type and graphics superior to a model with a 9-pin configuration. But some printers equipped with fewer pins can operate in a near-letter-quality (NLQ) mode. When set in this fashion, printed documents are almost letter quality in appearance. When switched back to the standard mode, the printing reverts to the normal dot-matrix format.

Besides the development of printers with a greater number of pins and an NLQ setting, the typical printer's output has been increased. Many models produce 180 or more characters per second in the standard draft mode, while other units are capable of printing at much higher rates.

The second major category of printers, the *laser printer*, is a relative newcomer to general computers. This device can produce documents and graphics that are near-typeset in quality.

As the name implies, a small laser is the heart of a printer's engine, the actual printing device. Other components include a photoconductor drum, toner particles, which make up the image much like the toner particles of a copy machine, and the paper.

The introduction of the laser printer in the general business and consumer markets helped trigger a publishing revolution. When a laser printer is combined with a computer and a desktop-publishing program, a new publishing tool emerges. This system is a compact publishing outfit that literally fits on a desktop. It has provided businesses, ranging from newspapers to small engineering firms, with the means to establish their own internal publishing departments. Even though the final copy may not be equal to that of commercially produced publications, it is nevertheless far superior to the output of a standard computer printer. It is, as stated, near-typeset in quality.

A laser printer does, however, have some deficiencies. It is more expensive than standard dot-matrix printers. But as is the case with computer hardware, the price is decreasing over time. A standard laser printer also has some limitations in terms of its graphics-reproduction capabilities, and this factor has implications for the way graphics and pictures are reproduced.

The third major type of printer, the ink-jet printer, literally uses a reservoir of ink to create letters, numbers, and graphics. Its output can almost match a laser printer in certain areas, and ink-jet printers are cost-effective. There are also compact models that are suitable companions for portable PCs.

The fourth category of printers includes the plotter, among other devices. A plotter uses a series of pens to create large-scale architectural and technical prints, as well as other line drawings. The output can range from plans for a sailboat to a new piece of machinery.

Finally, color printers are becoming increasingly popular. As described in Chapter 7, more software programs can handle color output, and a series of cost-effective printers has been marketed to meet a growing demand.

Local Area Networks

In brief, a *local area network* (LAN) is a communications system. It is confined to a limited physical area, such as an office or group of offices, and links PCs for the purpose of exchanging and sharing information and equipment. Printers, data-storage drives, and other devices are included on the network and can be shared by the network's users. This capability helps make a LAN an attractive option for an organization, allowing expensive high-speed printers and other equipment to support multiple users; these individuals can likewise share program and data files.

The design and implementation of a LAN is analogous, in certain respects, to a *multiuser system*, where a central computer is typically connected to a number of terminals. A terminal simply serves as a keyboard and a monitor, an interface device in this environment, since all the processing and storage manipulation tasks are performed by the central computer.[21]

A LAN, like a multiuser configuration, also makes it possible to share system resources. The LAN has an added advantage, though. Each PC has its own processing capability and may be able to operate independently if the network is rendered inoperable by an equipment failure. If the central computer "goes down" in the typical multiuser environment, the whole system comes to a crashing halt since the terminals are not capable of processing data on their own.

Even though multiuser systems are still widely used and are valuable in specific applications, they have been replaced by LANs in many cases. Cost-effective PCs and network-interface cards, as well as a LAN's decentralized processing capability, have contributed to this development. A LAN may also offer a solid, central administrative control over the computing environment, one of the traditional strengths of multiuser systems.

A LAN's topology, its physical layout, actually dictates how the different components that compose the LAN are linked together. The topology is usually based on a star, ring, or a bus design, and an important element in the overall configuration is the server. The server is a computer that manages and controls the flow of information through the network. It also stores program and data files.

In a typical operation, you are connected to the network by following certain log-on procedures, which may include using a password. At this point, you can gain access to the programs in the server.

Depending on the setup, you can use either the server or the PC you are working with to store your data files. When you finish for the day, you are disconnected from the network by using a log-off command.

As the data are relayed through a network to the various PCs, the nodes, the data must be routed to the correct destinations. In this respect, a LAN can be considered a super-highway overseen by a traffic cop. Since multiple PCs are interconnected, the data relays and requests must be organized to facilitate the information flow.

In the case of printing, a single printer typically serves multiple users. If several people select the print function on their word-processing program at the same time, the printer can print only one document at a time. So a queue must be used: the documents are printed, one after the other, based on the order in which the printing requests were received.

In addition to this approach toward networking, a *peer-to-peer* network can be created. Instead of using a server, PCs send and receive files to and from each other in this configuration. A peer-to-peer network also creates a more decentralized environment, and it can be less expensive to establish and operate.

The actual information flow in a network primarily takes place over one of three types of communications lines: *twisted-pair*, *coaxial*, or *fiber-optic* cable. Twisted-pair cable, which has the lowest information-carrying capacity, consists of two insulated wires that are twisted or entwined together. Coaxial cable is a su

perior relay line, and it is a shielded cable, a variation of which is commonly used in cable-television operations. A shielded cable is protected by a wire shield, which helps preserve the integrity of the relayed data since the cable, and consequently the information, is not susceptible to outside inter ference.

The last type of cable, fiber-optic cable, is the newest contender in this field. One of its major strengths is its large channel capacity. As described in a later chapter, an enormous volume of information in the form of light can be relayed on very thin glass fibers or threads.

Besides hardware developments, software manufacturers have introduced numerous products in support of LANs. These include word processors and other standard applications, as well as programs that take advantage of a LAN's interconnectivity to provide users with another way to exchange information. One such option is *electronic mail* (E-mail), in which messages are electronically relayed through the network to specified users. Rather than typing a memo, which must then be physically duplicated and delivered to one or more individuals, at the press of a button a user can electronically mail or deliver the same message via the network. This capability enhances the communication process within an organization and helps save time and money.[22]

Finally, it should be pointed out that the definition of a network extends beyond the configurations and applications described thus far. An example of a more novel use of a LAN, in this case as a network that has provided an integrated work environment, is the Manufacturing Automation Protocol (MAP). MAP is a manufacturing and industrial-based system. It has been used, in part, to automate various factory tasks and to link equipment produced by different manufacturers in a network.

Another example is the development of an electronic communications system on a national and international scale. LANs and individual PCs are islands unto themselves unless there is a mechanism to facilitate the exchange of information. A link that fulfills

Figure 3–10
PC software can be used in ergonomic design. The same program may also create images suitable for other projects (for example, desktop publishing). (Courtesy of Biomechanics Corp., HumanCad Division; Mannequin)

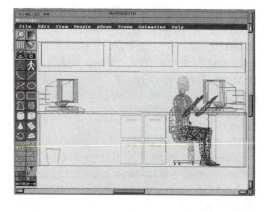

this goal is the *Internet*, a set of interconnected computer systems.[23]

By using the Internet as a data highway, you can communicate with colleagues and friends and gain direct access to information in other computers via your own system. It is a direct connection to resources that include library catalogs, electronic journals, software, and data files. In one example, images produced by NASA's spacecraft could be secured from the National Space Science Data Center.[24]

This potential to exchange information was enhanced in December 1991, when then-President George Bush signed the *High-Performance Computing Act of 1991*. The legislation's goal was the development of the *National Research and Education Network* (NREN), potentially a multi-gigabit data highway. The driving force behind the law was then-Senator Al Gore from Tennessee; the NREN could facilitate the flow of all types of information between researchers and educators.

In our information age, it's vital to have access to information in a timely manner. The high-speed links supported by the NREN could help make this possible. It is also hoped that the law would stimulate the development of advanced computing systems that a broader educational base could tap.

OTHER CONSIDERATIONS

This section covers other important computer-related topics that do not fit in the previous categories, including health and le-

gal issues. The chapter then concludes with a brief overview of an emerging set of computer applications that may have a profound effect on our lives.

Computers and the Work/Play Environment

While computers have altered our lives, they may be affecting our bodies in another way. There has been a growing controversy surrounding computer monitors. Monitors produce electromagnetic fields, electric and magnetic fields. The frequencies that concern us are *very low frequency* (VLF) and *extremely low frequency* (ELF) emissions. Much of the research has focused on the magnetic field, and a number of studies have been conducted to determine its effect on the human body. There are concerns that the emissions may play a role, for example, in miscarriages, birth defects, and cancer. The data, however, are conflicting and not conclusive at this time.

Herein lies the problem for computer users. The U.S. government has essentially ignored the issue, as of this writing, and the extent of the potential health problems is controversial. If the data are inconclusive, what can or should you do, if anything?

Until the data are conclusive, it may be wise to take some elementary precautions. You can buy a nonemission display or a monitor that conforms to Sweden's strict emission standards. You can also buy a special filter that fits in front of the screen. Despite advertising claims, though, the shields generally block the electric and not the magnetic field. Special monitor retrofit kits may also be available.

To further minimize potential exposure, the simplest precaution is to sit an arm's length from the monitor. This works as long as you don't move closer to someone behind you who may be working with another computer. Magnetic fields are stronger from the back and sides of monitors than the front. Finally, turn the monitor off when not in use.

In addition to the potential effects of electromagnetic fields, other safety and health factors should be considered. One of the most important is *ergonomic design,* the philosophy of developing equipment and systems around people, making equipment conform to an operator's needs and not the other way around. Desks, chairs, monitors, and keyboards are typical computer equipment influenced by ergonomic design.

Sound designs can help prevent some of the problems associated with working with computers. In one such affliction, *carpal tunnel syndrome,* your wrist can be damaged through repetitive keyboard motions. The problem can be alleviated or possibly even avoided by ergonomically sound keyboard and desk designs and by adopting good work habits. Take a break every hour and maintain a good sitting posture. A special wrist supporter placed at the bottom of the keyboard may also help.

Similar advice can help solve other problems, including eye fatigue and irritability. Don't place a monitor in direct line with a window since it might cause glare on the screen. A room should also be properly lit, in regard to light placement and intensity. And take a break every hour, to relax your eyes. A high-resolution display/board combination, which may also produce less flicker as a result of a higher screen refresh rate, can also help. The refresh rate has to do with the rate or speed in which an image is redrawn on the screen.

Besides taking care of ourselves, it's also important to protect information. As you probably know from news reports, the number of *computer viruses* is on the rise. A computer virus is a program that attaches itself, in a sense, to other programs. A virus may simply display a harmless message on a monitor or, depending on the virus, cause valuable data to be lost.

Viruses are spread from computer to computer by various means, including infected disks, some of which have been inadvertently shipped by commercial manufacturers. A virus may also remain dormant until triggered. In 1992 the well-publicized Michel angelo virus was triggered on Michelangelo's birthdate. Computers have internal clocks that keep track of the time and date. When the right date rolled around, if your system was infected, the virus could have become active.[25]

There are ways to protect yourself from a virus, such as regularly backing up computer files. A backup copy can be used to recover from a virus and, as a bonus, from a hard disk failure or breakdown. Another precautionary step is to use a special program that can scan your system for a virus. If one is found, it may be possible to remove the virus and repair any damage. Most virus-detection programs are also regularly updated to handle new viruses and variants.

Legal Issues

The spectacular growth of the computer industry and the integration of its products in all levels of society has raised a number of legal issues. Chapter 4, in part, examines optical data-storage systems and the copyright question; Chapter 8 looks at computer graphics, including their use in court. This section serves as an introduction to fundamental computer software issues.

Software Piracy. One of the challenges facing the computer industry is software piracy: the illegal copying and distribution of software. The United States is currently the world leader in software production, and international piracy adversely affects the companies that write the programs. This, in turn, has a detrimental impact on the overall economy, especially as we continue to move toward an information society.

The piracy problem is a matter of intellectual property rights. A panel convened to examine the situation suggested, in part, that the U.S. government should increase its "antipiracy efforts" and "strengthen the enforcement of intellectual property rights abroad and in the United States."[26]

The last sentence is a key one since pirating is not only an international issue. While there have been crackdowns, led by the Software Publishers Association (SPA), a group of software-related companies, piracy is still rampant in the United States.

It is an almost impossible situation that points out a dilemma of the information age. The same tools that can create an information commodity can be used to steal the commodity, in this case, software.[27] It is also an ironic situation. As discussed in Chapter 10, the almost universal nature of the PC has important First Amendment implications. This same universal nature makes enforcement very difficult, since it's so easy to copy a program.

Legislation can be enacted against piracy in an attempt to protect intellectual property rights, but how can this legislation be enforced? A few major "pirates" may be caught, but what of the hundreds or thousands of individuals who may copy software either for sale or, more typically, for personal use?

In response to this problem, manufacturers have taken different steps to combat piracy, including the introduction of software-protection schemes to prevent copying. An alternate protection method supports backups. But when the program is started, the owner is prompted to type in a specific word or number from a page in an accompanying manual or information sheet. Without this information, the program won't run.[28]

Another solution is to avoid copy protection altogether and to rely on a good product, at a fair price, with a good support policy. The goal is to prompt an individual to buy the software instead of illegally copying it.

Finally, the overall piracy issue is further aggravated by the language of licensing agreements and the widespread use of LANs. An agreement, which is made between the software company and the buyer, can vary from company to company. For LAN administrators, it may be harder to track and keep control of the software due to the number of potential users.

Borland International, one of the major software houses, had adopted a very clearcut licensing agreement for its products. It states, in part, that

This software is protected by both United States Copyright law and international treaty provisions. Therefore, you must treat this software *just like a book,* with the following single exception . . . to make archival copies . . . for the sole purpose of backing-up our software and protecting your investment.

By saying "just like a book," Borland means . . . this software may be used by any number of people and may be freely moved from one computer location to another, so long as there is *no possibility* of it being used at one location while it's being used at another. Just like a book that can't be read by two different people in two different places at the same time, neither can the software be used by two different people in two different places at the same time. (Unless, of course, Borland's copyright has been violated.)[29]

The language is not ambiguous and clearly points out the buyer's rights. It also clears up the confusion with some agreements as to whether or not you can use a single copy of a program with your home and office PCs, as long as they are not being used *simultaneously.*

To sum up, software piracy is a major problem, especially in an information age, and manufacturers have taken steps to combat it. However, more can be done.

As indicated by the panel convened to examine this issue, the U.S. government can increase its efforts in this area, as one of a number of possible options.[30] Education is another key factor. Intellectual property may not be viewed as "real property," such as a gold watch or money. A common perception is that you shouldn't steal money, but it's all right to copy a disk.

Part of this problem is philosophical. While the legal system may safeguard intellectual property, the philosophical element may have, in an educational sense, lacked focus. We have to recognize and accept the philosophical basis behind the idea of ownership before we follow the legal guidelines. Otherwise, the only preventive measures are legal, and if they cannot be fully enforced, as may be the case with widespread software copying, the situation will continue.

L Is for Lawsuit. The computer industry, like many others, is a fertile ground for lawsuits. The software arena has been especially hotly contested and rapidly changing; the focus has been on *patent* and *copyright* protections and violations.

In general, "a copyright provides long-term conditional ownership rights in a specific expression of an underlying concept, without protecting the concept itself. A patent provides relatively short-term . . . conditional ownership rights in an underlying concept (the patent's 'invention') without reference to the particular concept's embodiment."[31]

A 1981 Supreme Court decision, *Diamond v. Diehr,* actually opened the floodgate for software patents. They can provide broader and wider protection than a copyright and, as such, are valuable legal and financial commodities.[32]

One of the most publicized lawsuits, as of this writing, has been the dispute between Apple Computer and the Microsoft Corporation for copyright violations. Apple contended that Microsoft's Windows borrowed heavily from Apple's GUI in regard to the interface's "look and feel." Microsoft argued that overlapping windows and other display elements, the points of contention, were not protected by Apple's copyright. Even if they were, Microsoft argued, these elements were covered by a 1985 licensing agreement between the two companies.[33]

The issue came to a head in 1992. At that time, most of Apple's claims were thrown out by U.S. District Judge Vaughn Walker in a series of rulings. This included Apple's claim that Windows was "substantially similar to the look and feel of the Macintosh user interface."[34] If the ruling had gone the other way, Apple could have had a hammerlock on GUI rights and, possibly, future developments.

Finally, this dispute is only one of many current and future cases. Companies are filing more patent applications, and the hardware end of the computer industry has experienced its own share of lawsuits.

A pessimist might say the escalation of patent applications and potential litigation has a negative impact on the computer industry as a whole. If, for example, the developers of the first electronic spreadsheet had pursued and were granted a patent, they could have blocked the development of rival software products.[35] This, in turn, could have hampered the industry's growth. The patent process is also laborious and very expensive, and individuals and small businesses may be shut out from this form of protection.

An optimist might contend that legal protection can promote an industry's growth. There is a financial incentive for developers

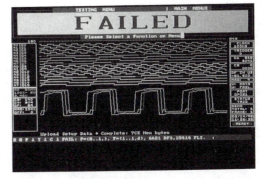

Figure 3–11
AI in the work environment. A screen shot from the ExperTest® general purpose tester. The ExperTest can be used as a test development bed for PC motherboards, among other applications. By drawing on AI technology, the system can deliver an enhanced diagnostic capability, much like a human expert. (Courtesy of Array Analysis)

to continue their work since they know their investments can be safeguarded and potentially rewarded on the open market.[36] Consequently, new products can be introduced, license agreements can be made with other companies, and the influx of new ideas in the marketplace may lead to other products, to the industry's benefit. It is, after all, a balancing act between protecting intellectual property while ensuring that a dynamic industry can continue to grow.

Artificial Intelligence and Related Topics

The last section of this chapter examines artificial intelligence (AI), another computer area relevant to our discussion of communications/information technologies and processing.

The AI field is dedicated, in part, to developing computer-based systems that seemingly duplicate the most important of human traits, the ability to think or reason. This section serves as a brief introduction to the field and related topics.[37] Detailed information about AI systems can be found in the references in the Additional Readings section of this chapter.

Natural-Language Processing. Natural-language processing focuses on simplifying the communication process between humans and computers. In one example, instead of responding to a series of commands that users have learned, the computer can learn and understand our native language. Consequently, for a database program, a query for information can be phrased in an ordinary sentence rather than a series of commands issued with keywords.[38]

This type of interface could be valuable for inexperienced users and those who don't want to learn a series of commands. It could also be extended to other programs to simplify different operations and make computer technology more accessible to a broader group.

Speech Recognition. Speech recognition, a subset of natural-language processing, allows a computer to recognize human speech or words. You directly instruct the computer to perform an operation instead of relying on the keyboard or other interface for inputting instructions.

During an operation, the system will initially digitize your voice, and then it must recognize various words before the assigned instructions can be carried out. There must also be a provision to ensure that certain words, such as *sea* and *see*, can be differentiated.

In one application, while working with a microscope, you could issue a command to a suitably equipped PC. Or instead of using a keyboard when using your favorite software, you could verbalize specific commands or functions. This same capability could also be incorporated into other devices, including consumer electronic products.

Regardless of the application, this type of operation simplifies your work and may help you simultaneously perform more than one task. It is also a very natural interface since it employs one of our most common communication tools, speech.

Finally, to complete the circle, the computer could answer you in turn. At MIT's Media Lab, one of the leading institutions in speech research and other fields, the "conversational desktop" has demonstrated this capability, and more. The system could recognize the speaker and could reply by stored speech, where "a number of phrases are recorded separately and then assembled into meaningful sentences as needed."[39]

At the PC level, a *speech-synthesis* card or board can be used for computer-generated speech. When combined with special software, a blind PC user can, for example, type on a keyboard, and the words are vocalized as they are created. Similar access is provided for information that appears on a screen, and a program may even be compatible with a GUI. In this case, menu selections are vocalized, and on-screen icons can be described.[40]

This capability can be linked with an optical character recognition (OCR) system, as examined in Chapter 7. An OCR configuration can recognize text from publications and other printed documents. The information can subsequently be reproduced in computer-generated speech, thus making the material available to visually impaired individuals.[41]

When this capability is viewed with the entire spectrum of speech research, it's an important achievement. Even at this developmental stage, our productivity is enhanced, the human–computer interface becomes increasingly transparent, and working with computers turns into a very natural process. More pointedly, these tools can help individuals to communicate their thoughts and ideas in a manner that would have been impossible to achieve a few short years ago.

Expert Systems. An expert system is a computer-based advisor. It is a computer program that can help in, among other areas, the medical and manufacturing fields.

The heart of an expert system is knowledge derived from human experts and includes "rules of thumb," or the experts' real-world experiences. Additional information sources could encompass books and other printed or electronic documents.

This expertise can then be retrieved during a consultation session. The computer leads you through the session by posing different questions. By using your answers, in conjunction with a series of internal rules, essentially the stored knowledge, the computer eventually reaches a decision.[42]

In one application, an expert system was designed to help service bureaus better define the potential target audiences for their clients, companies that primarily engaged in direct marketing sales through catalogs and mail-order transactions. In this operation, the program could compile a mailing list that matched a company's products and services with likely customers.[43]

In another application, the Foundation Bergonie, a French research center, developed a system that would "help doctors in general hospitals in the management of breast cancer patients."[44] The expert system provided the doctors with a level of advice that wouldn't, in normal circumstances, be readily available. In this situation, a computer program, which may have been viewed by some as just a collection of facts, could have a direct and beneficial impact on patient care.

In other examples, expert systems have been used to promote aquaculture, as described in Chapter 4, and by credit-card companies and the government. Consequently, expert systems have emerged as valuable information tools. They can

- help fill an information gap, as was the case with cancer patients where doctors were able to draw upon the knowledge and expertise of their colleagues.
- support a field where there may be too few human experts.
- preserve, in a sense, the knowledge and experience that would otherwise be lost when a human expert dies.

Yet, despite their advantages, expert systems are not infallible. Their capabilities are limited by the quality of the information, the rules that may govern the system, and by other criteria. If, for instance, a novel situation is encountered, such as a new physical reaction by a cancer patient, an expert system may not be able to support the physician working on the case to any significant degree. A human expert, in contrast, could more readily adapt to this new set of conditions.

There is also the problem of working with a machine. A human expert is typically better equipped to ask the right questions to define a client's situation. This may enable the human expert to arrive at a superior solution.

Another potential trouble area is litigation. If an expert system makes a mistake, who is liable? Depending upon the circumstances, is it the developer, the expert, or the individual who used the system, if it was designed to function only in an advisory capacity?

These questions, the high cost for liability insurance, and other factors, have held up the release of medical expert systems.[45]

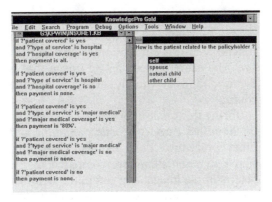

Figure 3–12
A screen shot of an expert system. The window on the left shows the code written to develop such a system. The window on the right shows one element of this expert system in action: a question and list of possible answers. (KnowledgePro screen shot by permission of Knowledge Garden, Inc.; KnowledgePro for Windows)

Other fields could be similarly affected, and in fact, in an article devoted to this topic, two pages outline potential risk areas for expert-system developers and designers.[46]

Finally, expert systems have been joined by another, related field, *neural networks*. In brief, "neural network technology, which attempts to simulate electronically the way the brain processes information through its network of interconnecting neurons, is used to solve tasks that have stymied traditional computing approaches."[47]

This capability is a reflection of a neural network's operational parameters. A network can, for example, be trained: it learns by example. A neural network can also work with incomplete data, much like humans, but unlike traditional computer systems.[48]

These characteristics make a neural network well suited for a variety of tasks, including optical character recognition. In this case, a system could recognize characters that were degraded and not perfectly defined.[49] In other examples, neural networks have been and could be trained to handicap horse races, to make stock market predictions, to adjust a telescope to improve its performance, or to recognize a face, even if the expression changes.[50] Similarly, and in keeping with the communications field, research has even been conducted to employ a neural network as part of a sophisticated tool for measuring television audiences.

Computer Vision. Computer vision can be described as the field in which a picture is fed to and analyzed by a computer, the machine equivalent of the human brain. Software is integrated with hardware components, including a videocamera, which functions in this configuration as a human eye. The camera produces an image that is digitized and eventually analyzed by the computer.

During this process, objects in a scene can be identified through *template matching* and other procedures.[51] In template matching, stored representations are compared to the objects in the image. The computer identifies the various objects when the templates match.

The capability to recognize different objects is a powerful feature since it can be harnessed to create machine and robotic vision systems. The information produced by a videocamera can, in one operation, be processed to inspect circuit boards for defects. Depending upon the setup, if a board is found to be defective, it could be removed from the assembly line by a machine controlled by a computer.

A more advanced computer-vision configuration, which also highlights a sophisticated AI application, is the development of an autonomous robotic vehicle. An autonomous vehicle would likewise process visual information and, in this situation, could function without direct human intervention.

An example of an ambitious project in this area is an autonomous rover equipped with a vision system for the exploration of the solar system. This vehicle would be activated upon its arrival on Mars or some other world. It would subsequently explore this body's surface, conduct experiments, and relay pictures back to Earth.

An autonomous vehicle would be an attractive exploratory probe, especially as we move farther away from the Earth in the course of planetary missions. As the distance from the Earth increases, the link to the probe would become more tenuous in view of the time it would take for a radio signal to make the round-trip between both worlds. This time factor could be critical.

In one situation, if a vehicle is traveling in a region littered with craters and other obstacles, and the pictures reveal a deep crater in the vehicle's path, it might be too late for the operators on Earth to take corrective actions. An autonomous rover would help

eliminate this problem since the vision system would provide the on-board computer with information about the surrounding terrain. Based upon an analysis and interpretation of the information, the computer would identify the crater and could issue an appropriate course change to the rover.[52]

This vehicle would be born from the union of various hardware components and a very sophisticated AI program that could process an enormous volume of picture information. The integration of computer software and hardware may also serve as a model for vehicles built for use on Earth. These include experimental vehicles that already exist and those that may be placed in production.

Philosophical Implications.

I. A robot may not injure a human being, or through inaction, allow a human being to come to harm.
II. A robot must obey the orders given it by human beings, except where such orders would conflict with the first law.
III. A robot must protect its own existence, as long as such protection does not conflict with the first or second law.
—Isaac Asimov, *The Robots of Dawn* [53]

The development and integration of AI systems in society has a number of philosophical implications. Some AI opponents believe, for example, that the new generation of intelligent machines will eliminate millions of jobs. There is also the fear that AI systems, including robots, diminish us as humans: we may lose a part of our humanity. Other AI-based systems have also raised the spectre of machines wreaking havoc, as presented in such movies as *The Terminator, Creation of the Humanoids,* and *Colossus: The Forbin Project.*

While these concerns may be legitimate, AI proponents have offered a number of counterarguments.

1. The tools of the AI field are just that, tools. An autonomous vehicle will not, for example, necessarily replace the human exploration of the planets, just as expert systems did not replace doctors.

2. While AI systems could lead to mass job displacement, it could be argued that the

technologies are not at fault. Rather, we have failed to cushion the impact of this development through job-training programs and other actions, and as stated, AI products such as expert systems are generally designed to complement and not replace humans.

3. Instead of diminishing our humanity, AI technology may have presented us with a set of tools that can enhance our lives. In one case, an expert system has improved a cancer patient's level of care, and in another, speech-recognition and -synthesis systems have made it possible for individuals to have better control over their environments.

4. The last concern was addressed, in part, by Isaac Asimov in his "Three Laws of Robotics," which opened this section of the chapter. The laws cut across scientific and science-fiction boundaries and could ultimately guide us as we start implementing advanced AI products.

Finally, to wrap up this discussion, it's important to remember that while the arguments presented by both camps may have some merit, we are ultimately responsible for AI systems, their role in society, and their impact in our lives. If used correctly, AI and other technologies may continue to enhance our exploration, the human exploration of our own world, of other worlds, and, just as important, the exploration of the human condition.

CONCLUSION

Computer technology will exert an even greater influence upon the communications field, and inevitably upon society, as more sophisticated computers are purchased by an expanding user base. Prices will continue to fall, and a combination of these factors, advanced technology at a lower cost, will eventually contribute to the remaking of the computer world, especially at the PC level.

In this regard, the late 1980s and early 1990s witnessed an important development in the family of workstations, the specialized computers briefly mentioned earlier in this chapter. Computers pioneered by companies such as Sun Microsystems and Apollo Com-

Figure 3–13
Joel and the "Bots," from the popular television show "Mystery Science Theater 3000." (Courtesy of Best Brains, Inc., 1992) [photo credit: Michael Kienitz]

puters offered users the processing power of a minicomputer combined with the graphics capability of a dedicated graphics station at a much lower cost. The computers also sported sophisticated multitasking and windowing capabilities, and could be networked.

These same features have been and are still trickling down to the general PC market. Personal computers, including the top models of the Apple and IBM PC lines, have emerged as workstations in their own right even though they may not be as sophisticated as the high-end units. The processing, graphics capabilities, and software sophistication of this new generation of computers represent a substantial design leap over earlier models.

This refinement of technology in the PC world has progressed at a dizzying pace. Eventually, the characteristics that separate the different classifications of computers will blur to the point where it may be difficult to distinguish one family of computers from the other. At this time, the PC will emerge as a small desktop or even portable computer that will match the capabilities of computers that were once the exclusive domain of major engineering and design firms.

As it has in the past, computer technology will continue to influence the way we learn, think, and communicate within a business setting, in school, and in our own personal lives. This influence has positive and negative connotations, both of which are discussed in later chapters.

REFERENCES/NOTES

1. Anthony Ralston and Edwin D. Reilly, Jr., eds., *Encyclopedia of Computer Science and Engineering* (New York: Van Nostrand Reinhold Company, 1983), 969.
2. Other computer systems are discussed when appropriate. These include mainframe and minicomputers in a later section of the chapter.
3. Alan Freedman, *The Computer Glossary* (New York: AMACOM, 1991), 301.
4. "Glossary," *Publish* 6 (October 1991): 124.
5. Advertisement, *Byte* 7 (December 1982): 583.
6. Sales catalog, Damark International, Inc., 1992, C1.
7. Jon Udell, "Windows Meets the Pen," *Byte* 17 (June 1992): 159.
8. Laurie Flynn, "Is Pen Computing for Real?" *InfoWorld* 13 (November 11, 1991): 75. Please note: Think about your own writing style and its particular characteristics. You might even have trouble recognizing your own writing on a note at a later date.
9. MS-DOS is Microsoft's DOS product for IBM PCs. It should also be noted that a PC may be able to support advanced functions, such as multitasking. Multitasking enables a computer to run more than one program simultaneously and can make an individual's computer work session more productive. This feature was not originally native to MS-DOS, in contrast to a PC where this function may have already been fully integrated in the system. Furthermore, other operating systems, such as OS/2 and UNIX, provide other powerful capabilities. UNIX, in fact, is well established, is gaining broader industry-wide support, and is available for a range of computer systems.
10. Frank Hayes and Nick Baran, "A Guide to GUIs," *Byte* 14 (July 1989): 250.
11. X-Window is an industry-wide system that is not tied to any particular platform. See David Moore, "The Migration of the X Window System," *Byte, IBM Special Edition* (Fall 1990): 183–185. Presentation Manager and Windows have been used with the OS/2 and MS-DOS operating systems, respectively. Windows, in a different version, is an operating system in its own right.
12. Betty A. Toole, "Ada, Enchantress of Numbers," *Defense Science and Electronics*, Spring 1991, 32. Please note: This issue has a number of articles about the Ada programming language and can be used to look at the language's roots.
13. Gene Apperson and Rick Doherty, "Displaying Images," in *CD-ROM Optical Publishing*, vol. 2 (Redmond, Wash.: Microsoft Press, 1986), 134.
14. Corel Systems Corporation, *Technical Reference* (Ottawa, Ont.: Corel Systems Corporation, 1990), 18.
15. At times, it may be advantageous to convert a bitmapped image into a vector-based one. This can be accomplished with different software modules and packages, and the quality of the conversion can vary, depending on the program and the original image's complexity. There is also some cross-compatibility between software in that a drawing program may, for instance, support bitmapped graphics.
16. Impulse, Inc., *Turbo Silver 3.0 User Manual*, 1988, 17.
17. For a look at single-frame controllers, please see Lou Wallace, "Precision Control," *Amiga World*, Special Issue 1992, 39–42. This particular article covers Amigas.
18. A program may also support audio and other multimedia data. See Chapter 8 for a description of multimedia systems.
19. Phil Neray, "Visualizing the World and Beyond," *Photonics Spectra* 25 (March 1991): 93.
20. Jim Martin, "Supercomputing: Visualization and the Integration of Graphics," *Defense Science*, August 1989, 32.
21. Multiuser systems are powered by a powerful PC or, more typically, a mainframe or a minicomputer. A mainframe can be categorized as a physically large, powerful, and expensive computer that can accommodate numerous users through a multiuser environment. A mainframe is equipped with enhanced processing and memory systems physically built into a central console, the "mainframe." A minicomputer can be considered a scaled-down version of a mainframe. It can also support a multiuser environment, as can suitably equipped PCs, albeit to a lesser degree than a mainframe.

22. E-mail and other products, which allow more than one person to collaborate on a project, are part of a growing software category called *groupware*.

23. The Internet was nominally geared for educational and research applications. But business uses are on the rise.

24. Space Digest, Bitnet, Ron Baalke, 10/31/91, downloaded information.

25. "Information Sheet on the Michelangelo Virus," compiled by J. M. Allen Creations/Michael A. Hotz, February 17, 1992.

26. Gary M. Hoffman, *Curbing International Piracy of Intellectual Property* (Washington, D.C.: The Annenberg Washington Program, 1989), 7. Please note: The panel was not solely concerned with software issues. Other pertinent topics included the pirating of videotapes and movies.

27. *Ibid.,* 10.

28. Sim City, a computer simulation program that provides you with the tools to build a city and the supporting infrastructure, has used a creative variation of this protection scheme. If you don't respond correctly to an opening prompt, you can still play the simulation, but your city is hit with different natural disasters, including a series of earthquakes that move and shake the on-screen landscape.

29. Borland International, *Turbo Prolog Reference Guide,* (Scotts Valley, Calif.: Borland International, 1988), C2.

30. See pp. 19–24 of *Curbing International Piracy of Intellectual Property* for a detailed discussion of the panel's recommendations.

31. Steve Gibson, "U.S. Patent Office's Softening Opens Floodgates for Lawsuits," *InfoWorld* 14 (August 17, 1992): 36.

32. Thanks to A. Jason Mirabito, a partner in the patent law firm of Wolf, Greenfield, and Sacks, for his suggestions for this section.

33. Beth Freedman, "Look-and-Feel Lawsuit Expected to Go to Trial," *PC Week* 9 (February 24, 1992): 168. Please note: The case also involved the Hewlett-Packard Company and its NewWave product.

34. Jane Morrissey, "Ruling Dashes Apple's Interface Hopes," *PC Week* 9 (August 17, 1992): 117.

35. Brian Kahin, "Software Patents: Franchising the Information Infrastructure," *Change* 21 (May/June 1989): 24. This is a sidebar in Steven W. Gilbert and Peter Lyman, "Intellectual Property in the Information Age," *Change* 21 (May/June 1989): 23–28.

36. "Patents: Protecting Intellectual Property," *OE Reports* 95 (November 1991): 1. Please note: This interview with a patent lawyer provides a good overview of patent law and what can and cannot be patented.

37. See Barry A. McConnell and Nancy J. McConnell, "A Starter's Guide to Artificial Intelligence," *Collegiate Microcomputer* 6 (August 1988): 243, for an introduction to AI's important elements.

38. The Q&A program used such an interface. It also had the capability to add words to its vocabulary.

39. Stewart Brand, *The Media Lab* (New York: Penguin Books, 1988), 51.

40. Richard S. Schwerdtfeger, "Making the GUI Talk," *Byte* 16 (December 1991): 118. Please note: Even though speech synthesis may not be considered an AI element, it is included here for organizational purposes.

41. Joseph J. Lazzaro, "Opening Doors for the Disabled," *Byte* 15 (August 1990): 258. Please note: The article also provides an excellent overview of PC-based systems for the blind, deaf, and motor disabled.

42. The representation of knowledge by rules is only one technique that can be used in the development of an expert system. By the way, an expert system can be created with a conventional programming language and with special programs called expert system shells. A rule can take the form of:

IF the car doesn't start AND
the lights don't turn on
THEN the battery needs charging.

This example is very simplistic, and in a real-world situation, multiple rules would be employed.

43. Persoft Inc.,"Separating Fact from Fiction about More/2," *News from Persoft,* press release.

44. Texas Instruments, "French Expert System Aids in Cancer Treatment," *Personal Consultant Series Applications,* product information release.

45. Edward Warner, "Expert Systems and the Law," *High Technology Business,* October 1988, 32.

46. G. Steven Tuthill, "Legal Liabilities and Expert Systems," *AI Expert* 6 (March 1991): 46–47.

47. Michael G. Buffa, "Neural Network Technology Comes to Imaging," *Advanced Imaging* 3 (November 1988): 47.

48. Gary Entsminger, "Neural-Networking Creativity," *AI Expert* 6 (May 1991): 19.

49. Larry Schmitt, "Neural Networks for OCR," *Photonics Spectra* 25 (October 1990): 114.

50. Maureen Caudill, *Neural Networks Primer* (San Francisco: Miller Freeman Publications, 1989), 4. Also see Andrew Stevenson, "Bookshelf," *PC AI* 7 (March/April 1993): 30, 38, 57, 58, for an in-depth review of *In Our Own Image*, by Maureen Caudill, a book relevant to this overall discussion.

51. Louis E. Frenzel, Jr., *Crash Course in Artificial Intelligence and Expert Systems* (Indianapolis, Ind.: Howard W. Sams & Co., 1987), 201.

52. A neural network could be very appropriate in this type of situation, especially since it could be trained to avoid numerous obstacles and deal with novel situations.

53. Isaac Asimov, *The Robots of Dawn* (New York: Doubleday and Company, Inc., 1983), back cover.

ADDITIONAL READINGS

Benfer, Robert A., and Louanna Furbee. "Knowledge Acquisition in the Peruvian Andes." *AI Expert* 6 (November 1991): 22–27.

Keyes, Jessica. "AI in the Big Six." *AI Expert* 5 (May 1990): 37–42. And, Shafer, Dan. "Ask the Expert." PC AI 3 (November/December 1989): 40, 49. The first two articles examine different expert-system applications; the third examines some of the differences between expert systems and neural networks.

Blackwell, Mike, and Susan Verrecchia. "Mobile Robot." *Advanced Imaging* 2 (November 1987): A18–A21.

Caudill, Maureen. "Driving Solo." *AI Expert* 6 (September 1991): 26–30. A look at autonomous vehicles.

Van De Bogart, Willard. "The Apparel Industry and Imaging." *Advanced Imaging* 4 (April 1989): 30–34, 75. An examination of CAD software, in addition to other technologies, in the apparel industry.

Brody, Herb. "The Great Equalizer." *PC Computing*, July 1989, 93. A comprehensive look at how PCs can "empower the disabled."

Busse, Torsten. "Software Floods the Patent Office." *InfoWorld* 13 (September 30, 1991): 39, 42, 44. An in-depth look at copyrights, patents, and computer software.

Byte 18 (April 1993). This issue of *Byte* focuses on visualization. It includes articles such as "Overview: Visualization: Seeing Is Believing," 120–28, by Jack Weber, and a list of visualization software.

Conrad, Fran. "What Should We Do? VDTs, Radiation, and Reproductive Risks." *Monitor*, July-December 1989, 11–15. The article examines potential risks from and provides background information about monitors.

Freedman, Alan. *The Computer Glossary*. New York: AMACOM, 1991. As stated in the text, this book is an excellent and comprehensive reference. It covers the computer field, its terminology, and the relationships between its component parts.

Frenzel, Louis E., Jr. *Crash Course in Artificial Intelligence and Expert Systems*. Indianapolis, Ind.: Howard W. Sams & Co., 1987. An excellent guide to AI technology and applications.

Gibson, Steve. "An Old Copyright Principle Sank Apple's Suit Against Microsoft." *InfoWorld* 14 (September 7, 1992): 32. And, Morrissey, Jane. "Copyright Lawsuit Enters Fifth Year with No End Near." *PC Week* 9 (August 17, 1992): S/5, S/9. And, both articles provide a good overview of the chronology of the Apple/Microsoft lawsuit and an analysis of why and how the eventual ruling evolved.

Glassner, Andrew S. "Ray Tracing for Realism." *Byte* 15 (December 1990): 263–71. A detailed examination of ray tracing.

Godnig, Edward G., and John S. Hacunda. *Computers and Visual Stress*. Charlestown, R.I.: Seacoast Information Services, Inc., 1990. A guide about computers and visual stress, this book also has a series of visual-training exercises, along with advice on how to deal with this problem.

Holtzman, Jeff. "Touch Screen Technology." *Hands-On Electronics*, April 1987, 34–36. A look at different touch-screen technologies and how they work.

Kehoe, Brendan P. *Zen and the Art of the Internet*, rev. 1.0. February 2, 1991. A reference guide to the Internet and some of the amazing information resources and organizations you can access, ranging from scientific data to community-based systems.

Kincaid, John, and Patrick McGowan. "When Faced with the Office Wiring Decision, 'Let the Buyer Beware.'" *Communication News*, February 1991, 59–61. And, Quraishi, Jim. "The Technology of Connectivity." *Computer Shopper* 11 (May 1991): 187–98. And, Rosch, Winn L. "Net Gain." *Computer Shopper* 12 (April 1992): 534, 536–44. Overviews of LAN technology. The first article focuses on the different types of cabling used in a LAN and their advantages and disadvantages; the third includes separate sidebars, such as a glossary of networking terms and a guide to buying a network.

LaQuey, Tracy. "Networks for Academics." *Academic Computing*, November 1989, 32–34, 39, 65. An overview of different computer networks, including the Internet.

Noakes, Robert. "An Introduction to How Data Communications Works by Networking of Computers, Terminals, and Modems." *Communications News*, March 1986, 31–36. An excellent overview of how a modem works in relation to a communications link.

Pournelle, Jerry. User's column. *Byte*. An ongoing column in *Byte*. Jerry Pournelle is a well-known science-fiction writer and computer expert. His monthly column covers a range of topics for the computer user, especially how hardware/software may work in real-life situations. It is also a user-friendly guide to applications and future developments. As Pournelle might say, this column is "highly recommended."

Public Service Center, U.S. Patents and Trademark Office. "Patents in Brief." Information sheet, February 1991, Washington, D.C. Downloaded from CompuServe, an on-line information service. An overview of patents and filing procedures.

Rivlin, Robert. *The Algorithmic Image*. Redmond, Wash.: Microsoft Press, 1986. A history of computer graphics and applications, richly illustrated with black-and-white and color images.

Smarte, Gene, and Andrew Reinhardt. "1975–1990: 15 Years of Bits, Bytes, and Other Great Moments." *Byte* 15 (September 1990): 369–400. An interesting history of *Byte* magazine that depicts the magazine's progression alongside the development of the microcomputing industry, as well as major world events. Excellent illustrations appear throughout the piece.

Sonera Technologies. *DisplayMate Video Display Utilities Reference Manual*. Rumson, N.J.: Sonera Technologies, 1990. The reference manual for Sonera Technologies' software is an excellent source of information about PC monitors. The software, the primary product, allows you to run your monitor through a series of evaluative tests. The accompanying manual extends far beyond the traditional software manual and is, in fact, a tutorial about monitors and their operation.

Stanley, Jeannette, and Sylvia Luedeking, eds. *Introduction to Neural Networks*. Sierra Madre, Calif.: California Scientific Software, 1989. An easy-to-understand introduction to neural-network systems by the same company that produced a neural-network software program.

United States Code Annotated, *Copyrights*. St. Paul, Minn.: West Publishing Co., 1991. Includes coverage of the Protection of Semiconductor Chip Products.

Webster, Ed, and Ron Jones. "Computer-Aided Design in Facilities and System Integration." *SMPTE Journal* 98 (May 1989): 378–84. The use of CAD software in the television industry for different design applications.

Werth, Larry. "Automated Vision Sensing in Electronic Hardware." *Sensors*, December 1986, 16–25. An introduction to machine vision and relevant applications.

Winston, Patrick Henry, and Berthold Klaus Paul Horn. *LISP*. Reading, Mass.: Addison-Wesley Publishing Co., 1988. The guide to LISP and programming in this language of the artificial intelligence community.

GLOSSARY

Artificial intelligence: The field dedicated, in part, to developing machines that can seemingly duplicate the most important human trait, the ability to think or reason.

Computer workstation: A computer that combines the power of a minicomputer with the graphics capabilities of a dedicated graphics station.

Computer vision: The field that duplicates human vision through a computer and a videocamera system.

Disk drive: A mass data-storage device for computers.

Electronic mail (E-mail): Electronic messages that can be relayed (for example, over a LAN).

Expert systems and neural networks: An expert system manipulates knowledge rather than numbers. In essence, it can serve as an in-house expert in a specific field. It combines "book information" with the knowledge, skills, and experience of human experts. A neural network, for its part, simulates the way the brain processes information through its network of interconnecting neurons.

Graphical user interface (GUI): A visual, rather than text-based, interface.

Graphics programs: The generic classification for different categories of computer graphics software. These range from a paint program, where individual pixels are manipulated, to a draw program, where individual geometric objects or shapes are manipulated.

Human–computer interface: The tools we use to work with computers. They range from keyboards to touch-screens.

Local area network (LAN): A dedicated data communications network that links computers as well as peripherals, such as printers, for the purpose of exchanging as well as sharing information, programs, and other resources.

Monitor: A computer's display component.

Manufacturing Automation Protocol (MAP): An industrial-based network that links equipment and can be used to automate various tasks.

Multiuser system: A computer that can accommodate multiple users, typically through terminals. The computer provides the processing and data-storage capabilities, while the terminals serve as the interface between the users and the computer.

Natural-language processing: Natural-language processing focuses on simplifying the human–computer interface. Rather than using special keywords to initiate various computer functions, a user could accomplish the same tasks with simple word sequences.

Personal computer (PC): A computer typically designed to serve one user.

Programming language: One of the computer languages used to create a computer program, the instructions that drive a computer to complete various tasks. Typical languages include C, Pascal, and Fortran.

Printer: A device that produces a hard copy of the information stored by the computer.

Random-access memory (RAM): A computer's working memory.

4 Information Storage, Retrieval, and Communication

The 1980s and 1990s witnessed the growth of new transmission and information-storage systems that employ light to relay and store information, respectively. The information can range from a person's voice to pictures produced by a videocamera. This chapter examines the latter light-based system, optical-disc technology and data-storage configurations.

A single optical disc can store an immense volume of information. It also provides an effective tool to handle the mountains of data we must contend with daily.

Finally, the chapter concludes with an overview of hypertext and hypermedia programs. This class of software often enables us to organize information in a new way and, more important, communicate more effectively.

INTRODUCTION

The optical disc emerged as a major information storage tool in the 1980s. A popular application is the CD, an audio-storage system you may already own. A CD is a small, round disc that stores digital audio information in the form of microscopic pits. A CD is composed of multiple layers, one of which serves as an information layer containing the code of pits. Light produced by a laser initially writes or stores this information on the original master disc, which in turn plays a key role in the production of the commercially sold copies. A laser subsequently recovers or reads back this information when one of the copies, the CD, is placed in a player.

During a read operation, a beam of light generated by the laser scans the CD. The beam is then reflected to different degrees, in terms of the beam's strength, when it passes over the pits and unpitted areas called lands. The reflected light, an optical representation of the stored information, is picked up by a light-sensitive detector and converted into a corresponding electrical signal. The signal generated during this process eventually passes through a DAC, and the final output is an analog signal that is compatible with the analog world.

The CD and other members of the optical-disc family are constructed like sandwiches: these include the information layer, where information is stored, and the reflective metallic layer, which enables the read or playback operation.

The CD is only one part of a growing optical-disc field. After a brief overview of the optical-disc family, we will cover important characteristics of some of the major systems.

OPTICAL-DISC OVERVIEW

For our discussion, optical-disc systems fall into two categories: nonrecordable and recordable media. CD and conventional *compact disc read only memory* (CD-ROM) systems are nonrecordable. Write once, read many (WORM) and erasable systems are recordable.

Both categories of discs share several characteristics.

1. Information can be stored in the form of pits or, in erasable systems, through other techniques. This information is digital in nature, with the major exception of the videodisc.

Figure 4–1

Optical discs support numerous applications. Newer software releases also make it possible to tap this information resource, which in this case includes a premade CD control panel, possibly for a multimedia application. Multimedia is described in detail in Chapter 8. (Software courtesy of Asymetrix Corp.; Multimedia Tool-Book)

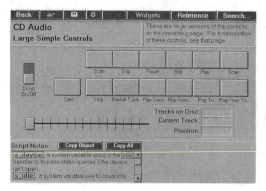

2. Optical discs are fairly rugged and durable, since a plastic coating protects the stored information from fingerprints and scratches. Also, the laser is focused beneath a disc's surface at the information layer, so dust and other minor surface obstructions may not adversely affect playback.

3. Unlike a conventional vinyl LP, a disc is not subject to wear, since the playback mechanism is not a stylus but a beam of light. The same disc can be played multiple times with no discernible loss in the signal quality.

Discs are not indestructible, however. For example, deep scratches can affect a CD or CD-ROM's playback characteristics, and the manufacturing method may affect a disc's longevity. Possible problem areas include the corrosion of a disc's metal layer.[1]

4. Optical discs can store megabytes of data, due in part to a laser's capability to distinguish between tightly recorded tracks of information. The different systems also incorporate sophisticated error-detection and checking schemes to ensure the integrity of the information. But error checking is much more critical for an optical disc that stores computer data than for a standard CD, due to the potential impact of the errors.

5. Like floppy and hard disks, optical discs are random-access devices. Unlike a tape-based system, which may require you to fast-forward through the entire tape to find a specific piece of data, optical discs provide almost immediate access to the information. This is a particularly attractive feature for numerous applications, including those used in a video environment.

6. The optical-storage field is rapidly expanding. New storage formats have been introduced since the early 1990s, and this trend will continue. Similarly, some of the systems covered in this chapter may be enhanced and sport new options, while others inevitably may fail and disappear from the market. Nevertheless, the list of applications should continue to grow.

SPECIFIC SYSTEMS

Compact Disc

A CD is a high-fidelity audio-storage medium, and its superb sound-reproduction qualities reflect the digital and optical recording and playback techniques. Interfering noise, which may affect a conventional LP record system, is greatly reduced in CD systems. A disc less than 5 inches in diameter can store approximately an hour of music.

A CD player is usually equipped with a sophisticated microprocessor that allows the user to quickly access any of the disc's tracks and select a predefined order of tracks for playback. These functions, in addition to a disc's small size, durability, and long playing time, have contributed to the CD's popularity with consumers and radio stations.

As described in "Standards" in Chapter 2, the CD industry is governed by an established set of standards, which makes all CD systems compatible. This factor played an instrumental role in its widespread acceptance.

The CD is, however, faced with competition from *digital audio tape* (DAT) and other digital tape systems (described in Chapter 9). A DAT player can record as well as play back digital tapes, and the audio quality is equal to that of a CD.

Compact Disc-Read Only Memory

A conventional CD-ROM, which looks like an audio CD, serves as a data-storage medium. A disc is preloaded with computer-compatible data, perhaps including computer programs.

A CD-ROM is interfaced with a computer via a CD-ROM drive and special driver software. Once the disc is placed in the drive, software bundled with the system provides the user with a mechanism to search and retrieve the disc's information.

The primary application of CD-ROM technology is electronic publishing. A disc can store more than 600 megabytes of a variety of data, including computer graphics and digitized pictures. This capacity makes the CD-ROM a very flexible medium with regard to the types of information it can accommodate.

CD-ROM Titles. An early CD-ROM release was the electronic text version of *Grolier's Academic American Encyclopedia*. This disc highlighted the CD-ROM's storage properties: an entire encyclopedia of some 30,000 articles was recorded on a single disc, with room to spare. It was also integrated in a PC environment, and information from the encyclopedia could be retrieved by word-processing programs. More recent encyclopedias on a disc have also incorporated sound, graphics, and animations.

Certain categories of information that can be retrieved through on-line database companies, which are typically reached via a telephone line and modem connection, have also been distributed on CD-ROMs. Other organizations that normally distribute their work in printed form have similarly adopted this electronic-publishing technology, typically to complement their hard-copy product lines.

Consequently, general and specialized encyclopedias, abstracts, indexes, and a range of information generated by government agencies have been recorded on CD-ROMs. The latter have included a number of discs available through the National Space Science Data Center. For a nominal fee, you can explore Venus or Mars from your armchair. The discs contain data generated from the Magellan and Viking Orbiter missions to Venus and Mars, respectively.[2]

CD-ROMs have also been used to distribute magazine articles. With a single disc, you can browse through literally hundreds of articles about computers, national and international news, or the business world. You can even buy a dedicated CD-ROM magazine. This electronic publication, originally produced for distribution via a disc, can include text, graphics, and music.

One of the more interesting of the first wave of discs was the PC-SIG CD-ROM. The PC-SIG is a source of public-domain and user-supported software (shareware) written for IBM PCs. The programs cover everything from computer languages to games, and the entire library could fill 1,000 or more floppy disks.

This disk library was transferred to a single CD-ROM, where individual programs could be retrieved. The application was and still is a valuable one for PC owners, since the CD-ROM may cost less than a hundred dollars, in contrast with a much higher price tag for the equivalent floppy-disk library. This economy helps make the CD-ROM an ideal distribution medium for large collections of computer programs.

The CD-ROM's cost-effectiveness and large capacity also lend themselves for efficient distribution of very complex software, such as a desktop-publishing package, along with a comprehensive computer-based training program and an extensive electronic *clip-art* collection. Clip art, in this context, is a collection of pictures a user can legally incorporate in a project. Clip art can include 2-D and 3-D graphics, as well as gray-scale and color images. The Corel Systems Corporation used a CD-ROM's capabilities to bundle a collection of several thousand clip-art images with Corel Draw, its high-end illustration program. The floppy-disk equivalent for the entire collection of information would have been approximately 500 disks.[3]

CD-ROMs are also valuable for libraries, especially if a library is facing storage and budgetary problems. CD-ROMs are cost-effective, can be used with a PC to search for specific information, and can save space. But the library must support the system with PCs.

Information Retrieval. The development of the CD-ROM and optical systems has presented us with an opportunity to gain access to entire libraries of information through a series of small discs. Yet, to tap this information effectively, suitable search methods must be devised.

A CD-ROM that serves as a database is equipped with software that functions as the information-retrieval mechanism, and depending on the package, it enables an individual to search through the information in different ways. Some systems employ keyword searches; others support more sophisticated mechanisms.

This searching capability highlights the power of the PC when combined with a mass-storage device. Instead of looking through different books and magazines to find a specific piece of information, you can let the PC do the work for you. If the data are stored on a disc, the retrieval process can be simplified and enhanced. This topic is discussed in further detail later in this chapter.

Other Issues. It is quite easy to become overly enthusiastic about CD-ROM technology and its applications. There are, however, some negative characteristics that should be discussed.

CD-ROMs are slower than hard drives, and a user must be sure that a given disc will work with his or her hardware system. Most drives support a read-only capability; recordable configurations, which could be used to master a disc, have been quite expensive.

Some of these issues have already been addressed by new generations of equipment. Faster CD-ROM drives have been manufactured, and less expensive, recordable CD-ROM systems (CD-R) are appearing on the market. Also, the price for conventional read-only drives has dropped to less than $250. While they may not incorporate the latest features, these inexpensive drives could make certain CD-ROM applications more accessible to a broader user base.

Besides these issues, a potential CD-ROM application by the Lotus Development Corporation raised a number of privacy questions. Lotus planned to take advantage of a CD-ROM's storage capabilities and sell a disc loaded with demographic data about American consumers. The company received so many complaints, though, that the product was dropped from its line.

This concern over privacy is not limited solely to the CD-ROM field. Similarly, in the early 1990s phone companies began supporting *Caller ID*. By selecting this option and using a special device, customers could identify a caller's phone number. This service sparked another privacy debate.

Proponents believed that Caller ID would reduce obscene and nuisance calls. It was also embraced by the business community since it could be used to identify customers. Opponents contended that Caller ID was an invasion of privacy. To resolve this issue, some phone companies initially agreed to provide callers with one or more blocking options to defeat the service. In some cases, a company may have selected this route in the face of possible legislative action.[4]

These examples highlight a key issue of the information age: an individual's right to privacy. While the new technologies can enhance our communications capabilities, they can also be invasive. The available options to protect yourself include consumer pressure, employing other emerging technologies, and adopting privacy regulations, as was the case with the European Community (EC).

The EC had proposed a series of strict regulations governing the collection and dissemination of personal information. The rules recalled the World War II era, when information from telephone records was used for political purposes. The proposed regulations, in contrast, would help protect an individual's privacy.[5]

While privacy is important, some people believed the regulations were too strict and would impede the flow of information between countries. A similar situation has prevailed in the United States. Some individuals, as well as government agencies, believe there has to be a balance between protecting an individual's right to privacy and the government's ability to gain access to specific types of information. This topic is discussed in Chapter 10.

Copyright. CD-ROMs and some of the other optical discs described in this chapter have raised a number of copyright questions. As is the case with software piracy, copyright holders of photographs and other material are concerned about the illegal distribution of their property. Once a photograph is on a CD-ROM, a user could retrieve and use the image without compensating the owner. This problem is only magnified in light of the CD-ROM's mass-storage capabilities. Now an individual could gain access to a whole library of images and sounds through a disc collection.

It's a complex issue. First, once the information is committed to an electronic form, it becomes widely susceptible to illegal copying.

Second, the new generation of image- and sound-manipulation software allow a user to retrieve an image or sound and alter it. Copyrighted material can be altered, and part or all of it can be used in another work.

Third, the rapidly expanding computer-based-presentation field is built on both still and moving images, including video clips; these images are distributed on CD-ROMs. There is a great demand for this raw material.

Fourth, simply keeping track of information is problematic. CD-ROMs and other optical media can accommodate megabytes of different types of information, and the material must be legally cleared for use. For example, if you are creating a compilation of pictures for distribution via CD-ROM, you must be sure the individual or organization providing a picture has the "full, clear, and unquestioned legal right" to the image.[6] If not, you could commit a copyright violation.

Fifth, enforcement raises serious questions. You can legislate against illegal copying, but how do you enforce the law? Are you going to differentiate between individuals, such as a hobbyist who unwittingly uses a copyrighted image in a newsletter and a producer who creates a contracted multimedia production? Or are they going to be treated equally? If an exception is made for the hobbyist, doesn't the copyright owner still pay the price, since he or she is not compensated for the work?

Sixth, the copyright issue must be examined from a system's perspective. As described in this and other chapters, the copyright question cuts across different fields. Possible violations are not the exclusive domain of the optical-disc industry, and consequently, the problem as well as possible solutions should be addressed systematically. For example, as discussed in Chapter 3, an

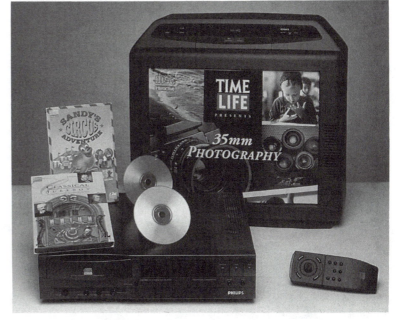

Figure 4–2
The Philips CD-I System with the Time-Life Photography *title on screen. (Courtesy of Philips Corp.)[photo credit: Richard Foertsh]*

educational campaign could be mounted to heighten awareness about copyright issues.

Compact Disc-Interactive and Other Compact Disc Formats

The *compact disc-interactive* (CD-I) and other emerging formats are members of the compact disc family. The CD-I is an optical-storage system primarily geared for the consumer and educational markets.

Unlike a conventional CD-ROM, the CD-I was originally designed as a stand-alone unit equipped with an internal computer.[7] The goal was to make a CD-I system attractive to consumers, since it was self-contained, simple to set up and operate, and its output was compatible with a standard television set.

The CD-I is a sophisticated audio-video tool that supports a variety of playback formats. In two of its audio modes, for instance, a disc can offer either CD-quality sound or one that would approximate the range of human speech. A disc can also support animations, video, and photographic-quality images.

The CD-I is also an interactive tool. A user interacts with the database of information

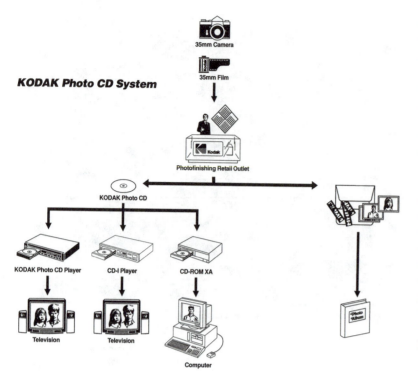

KODAK Photo CD System

Figure 4–3
The Kodak Photo CD System: the processing and playback options. This illustration also highlights an important characteristic of the optical storage field in general. The same tool can potentially support multiple applications. (Courtesy of Kodak Corp.)

through the CD-I player and a remote-control device. Instead of watching and listening to a prearranged presentation, much like a videotape, you select the information you want to hear and see and the order in which it is to be played. Two application areas include entertainment and education.

In the entertainment category, a CD-I game disc can present a wide range of choices. A flight simulator could be supported with realistic images depicting a variety of airports and scenery. Another type of entertainment disc can also function as an audio CD. In one example, a Louis Armstrong disc can be played on a CD player. But when used on a CD-I system, you can hear his music as well as view graphics and retrieve information about Armstrong's work.[8]

For educational applications, a CD-I system could provide a user with a sophisticated "how-to" course about topics ranging from automotive repairs to karate. Based upon the user's needs and requests, the computer would retrieve and display specific information stored on the disc. Actual CD-I titles

include tours of the Smithsonian Institute, where you can interact with the exhibits, and discs geared for children.

In addition to the CD-I, Commodore introduced its own interactive product, *Commodore Dynamic Total Vision* (CDTV). The CDTV system was essentially targeted toward the same market as the CD-I, and with accessories, this component could be used as an Amiga computer.[9]

Kodak also entered the compact-disc industry with the *Photo CD* system. As originally designed, up to 100 pictures could be transferred to and stored on a disc, as well as developed into slides and prints. A disc could then be played on a compatible CD-ROM drive or, for consumers, on a Photo CD player. The pictures in the latter configuration would subsequently be viewed on a television, and the same system could also play audio CDs.

The capability to play different discs is an important one and an industry-wide development. Instead of buying multiple players and/or PC-based drives, you can use one machine to accommodate various discs. This is particularly significant in view of the growing number of formats, including the CD, CD-I, Photo CD, CD + Graphics, the CDTV, and the CD-ROM XA.[10]

This capability also opens up new creative possibilities. By using the infrastructure that supports the Photo CD system, for example, you could create your own electronic, disc-based clip-art collection from original pictures. Store the images on a disc, and load the disc in an appropriate drive. The pictures could subsequently be retrieved and used through various interfaces, much like conventional graphics stored on a hard disk. This system would also be cost-effective since you wouldn't have to buy the processing equipment. Ultimately, you gain access to another creative tool for desktop publishing, desktop video, and multimedia applications.

This system, when viewed in context with the overall compact-disc family, is actually quite amazing. What started as a new audio delivery system, the audio CD, has blossomed into a series of discs that can accommodate a wide array of data and applications.

Game manufacturers, such as Nintendo, also promise to be powerful forces in this area.

Finally, the rapid growth of the optical-disc field highlights the growing convergence between different technologies and their respective applications. Products such as the CD-I marry computer-based applications with standard television technology. In this case, a television becomes a part of a computer-based entertainment and educational system. As discussed in a later chapter, the establishment of an enhanced and advanced television standard will only accelerate this trend.

Systems such as the Photo CD, for their part, cut across the traditional and silverless photographic fields. Pictures produced as standard prints or slides could also be stored on a disc and viewed on a television. The same pictures could subsequently be manipulated with a computer.

Videodisc

The videodisc is the pioneer product of the optical-disc family. The videodisc has been produced in different formats, including a discontinued nonoptical version manufactured by RCA.

The applications accommodated by videodiscs can be classified in two general areas. The first is dedicated toward the consumer; the second is geared for educational applications.

In the consumer category, videodiscs are used to distribute movies. Even though the VCR's popularity has affected the videodisc's market penetration, the disc has been well received by videophiles, those individuals who demand a superior-quality audio-video reproduction. A videodisc generally surpasses consumer-based VCRs in this respect, and the videodisc may also support a digital audio signal, even though the picture information is stored in analog form.

In the second category, a videodisc player can be interfaced with a PC, in one configuration, to create a sophisticated interactive environment. A computer program is designed for a specific task, and the software with the computer controls the videodisc player and retrieves the information from the videodisc. An interactive mathematics program illustrates this procedure.

Specific frames stored on the videodisc are retrieved based on a student's responses. If a problem is displayed on the screen and the student types an incorrect answer, the computer may select a series of frames that depict the appropriate problem-solving technique. If the student's answer is correct, a more difficult problem stored on another section of the disc could be accessed. The video display can also be overlaid with computer-generated text, and during the lesson, computer graphics and video images could be accommodated.

Besides this computer-dependent system, a videodisc player equipped with an internal microprocessor can run a less-sophisticated interactive program. The necessary instructions are stored on the disc in a more basic configuration, and a hand controller instead of a computer keyboard can serve as the input device.

Videodiscs have been adopted by libraries, researchers, and schools. In one application, discs have stored thousands of still photographs. A disc can subsequently be copied and distributed for much less cost than creating a comparable library composed of individual copies of all the original images. The same capability likewise extends to the CD-based optical field.

The videodisc and its player also make it possible to view a single picture for an extended time period, unlike a VCR. In this operation, the videodisc player rereads the same track where the image is stored.

A VCR can also provide a single-frame view. But the tape could be damaged if the VCR is paused for several or more minutes, since the same section of the tape would be in contact with the video heads.

WORM and Erasable Drives

An individual can write data to and retrieve data from a WORM disc, a permanent storage medium. Optical storage systems have been used for such operations for a number of years, including a 14-inch optical-disc

Figure 4–4
A screen shot from The Art of the Czars: St. Petersburg and the Treasures of the Hermitage, *a CD-I disc. An example of the rich pool of information electronic publishers have begun to tap and subsequently distribute. (Courtesy of Philips Corp.)[photo credit: Richard Foertsh]*

configuration introduced by the Storage Technology Corporation in 1983. The company's high-capacity disc could hold 4 gigabytes of data, the equivalent of two million pages of double-spaced text.[11]

Contemporary WORM systems, which can be interfaced with PCs, can hold well over 600 megabytes of data. A disc can also serve as an archival storage medium in view of its expected lifetime of 10 or more years. Ten years extends beyond the mandatory time period certain types of records must be retained.

In one application, a bank or other financial institution could use a WORM drive to create a permanent record of various financial transactions. This permanence is advantageous in certain circumstances. For example, since the information cannot be altered, a WORM disc could facilitate an audit at some future date.

A WORM configuration could also be used as an identification tool. A depositor's signature could be digitized and stored on a disc. When this individual conducts a transaction, the teller could immediately recall the stored signature for verification purposes.

In addition to this permanent storage system, erasable optical systems are becoming more popular. A number of companies have either produced or are in the process of developing such products.

In one design, the erasable discs are a component of a *magneto-optical* (MO) configuration that employs both optical and magnetic principles as the means to store and retrieve the data. A user can record and erase information with the same ease as with current floppy- and hard-disk systems, and information that is no longer pertinent can be eliminated to make room for new data. MO systems also have high data-storage capacities and are removable like WORM discs. Thus, a new disc can be used when the current one fills up.

Another type of erasable system is the *phase-change drive.* Phase-change drives also use lasers to read and write the data but, due to design differences, are faster than comparable MO systems.

Conclusion

Regardless of the type of optical drive or media, current and future optical systems provide us with the means to store large quantities of information. This is important: data-storage demands are increasing as the nature of information becomes more complex. In desktop video and multimedia applications, for example, 24-bit graphics as well as high-quality audio and video clips may be used in a project. While data-compression techniques can be applied, information of this nature is still data-storage intensive.

The different classes of optical systems will also continue to coexist in the marketplace. At first glance, a conventional CD-ROM system may appear to be obsolete in the face of some other configurations. But CD-ROMs are very cost-effective, may have a greater storage capacity than another type of optical system, and at the time of this writing, have the edge in market penetration.

Magnetic media, hard and floppy drives, will also continue to be used. Hard drives are faster than their recordable optical counterparts and are fairly compact in size, a critical design asset for portable PCs.

Fixed-storage hard drives are also less expensive than recordable optical drives, although this economic advantage decreases as the data-storage requirement increases. Basically, when dealing with mass-storage needs, it's cheaper to buy another optical disc than a new hard drive. While there are also removable hard drive systems, they don't match an optical disc's capacity.

Besides hard drives, the floppy disk will continue to play a central role as a distribution and storage medium. There is also a newer type of floppy, the *floptical*. A floptical disk uses a laser to track tightly packed data on a special magnetic medium. The same drive is also backwardly compatible with conventional floppy disks. This characteristic gives a user the best of both worlds, a mass-storage system of more than 20 megabytes per floptical and a drive that is compatible with the older, more established standards.[12]

Finally, as discussed in Chapter 2, the adoption of new technologies and their applications is typically evolutionary in form. This also holds true in the data-storage field, since optical systems have been gradually integrated in the current market structure.

In one instance an optical system, the CD, *eventually* supplanted another medium. CD-ROMs do not dominate the data-storage field, but they have become more popular over time as hardware costs decrease and the number of CD-ROM titles increases.

HYPERTEXT AND HYPERMEDIA

The megabytes and gigabytes of information that can be stored with contemporary optical and magnetic systems can pose a problem. How do you organize this information, and if applicable, how can you present or communicate it effectively?

As described in the CD-ROM section, software can be used as a retrieval tool. At this time, a promising information-retrieval system is *hypertext*, a word coined by Theodor Nelson, one of the pioneers of this field.

Operation

Hypertext is a sophisticated information-

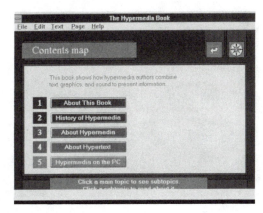

Figure 4–5
Screen shot of a hypermedia book. The numbered rectangular areas are used to activate links to gain access to other information. (Software courtesy of Asymetrix Corp.; Multimedia ToolBook)

management and -retrieval system that works in a nonlinear fashion and cuts across the magnetic and optical storage domains. The concept for such a system was originally conceived in the 1940s by Vannevar Bush, President Roosevelt's science advisor.

Hypertext operates in much the same way that humans think. You may, for example, use hypertext to conduct research about the new communications technologies via the information stored on a videodisc, CD-ROM, or other medium. While researching, you initially retrieve an article about fiber optics and see the term *laser diode*. Since the term represents an important bit of information, the words may be highlighted in our hypertext environment. This indicates that the words are linked to other information. If you move an on-screen cursor to a highlighted word and click a mouse button, the linked information is retrieved and displayed on the monitor, either on a separate page or in a window. When you finish reading, you return to the original page with another click of a button.

As you continue reading, the term *semiconductor* is similarly marked, and you could retrieve information about the semiconductor, and then, the semiconductor industry. These topics may, in turn, lead you to linked application areas and to overviews of the economic and social impact of the proliferation of these devices.

By following this pattern, you have retrieved information in a natural way, that is, the way people think. You follow a "train of thought" that enables you to make associations between diverse topics. In this context,

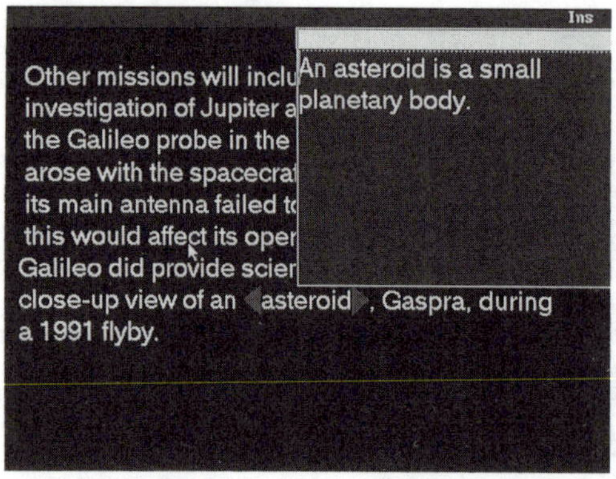

Figure 4–6
An example of a pop-up window. Note the link markers on either side of the word asteroid in the background text. (Software courtesy of Ntergaid, Inc.; Hyper-Writer!)

hypertext is no longer just a search mechanism. It provides a new way to organize, link, and communicate bodies of information and knowledge and, just as important, to see previously hidden connections between topics. A hypertext system performs these tasks effortlessly, at least for the user, and the various links can be retraced so an individual can return to a given source.

Hypermedia. We were initially introduced to hypertext programs on a wide-scale basis by *HyperCard*, a software product designed for Apple's Macintosh computer. HyperCard consists of blank, computer-generated but user-defined *cards*, much like a series of index cards, that can be used to create a *stack*. A stack is a group of related cards.[13] HyperCard supports information ranging from text to graphics, and a special programming language, HyperTalk, can be used for creating more sophisticated applications.

Besides HyperCard for Apple computers, IBM PC and other platforms have their own hypertext programs. *HyperWriter*, a program created by Ntergaid, Inc., is used, in the context of this book, to explore hypertext and hypermedia concepts and applications.

The term *hypermedia* has been associated with *multimedia*, a subject discussed in Chapter 8. Briefly, "multimedia is the integration of different media types into a single document. Multimedia productions can be composed of text, graphics, digitized sound," video, and other types of information. Hypermedia software, in turn, allows you to "form logical connections among the different media composing the document."[14] When you click on a highlighted term, you may see a picture or hear a sound, instead of simply a page of text.

With a hypermedia program, the links between the information, whether textual or graphic, can be represented in different ways. These include highlighted text, text placed between different symbols, and buttons. Buttons are visual markers, typically labeled, that can help create a more effective user interface. You click on a button and an action takes place, such as jumping to a new page of text.

You can activate a link by simply moving an on-screen cursor to the appropriate point in the document. Next, the link is activated by either clicking a mouse button or by a keystroke. As described, links are used to join the different *topics*, essentially the different information elements, in a document.[15]

HyperWriter supports different classes of links: those to and from text, graphics, and action. The text can be full pages of information or small pop-up windows. The pop-up windows are especially useful for short pieces of supporting data, as may be the case when defining a term mentioned in the body of the text.

Graphics links, as implied, tie information to a graphic. Action links, the last major category, serve various tasks. These include executing a different program and operating a videodisc player.

In reflection of its multimedia capabilities, the program also supports animations and digital sound. The sound capability is an important one in that it can enhance a document, much like audio can help a movie or television program.

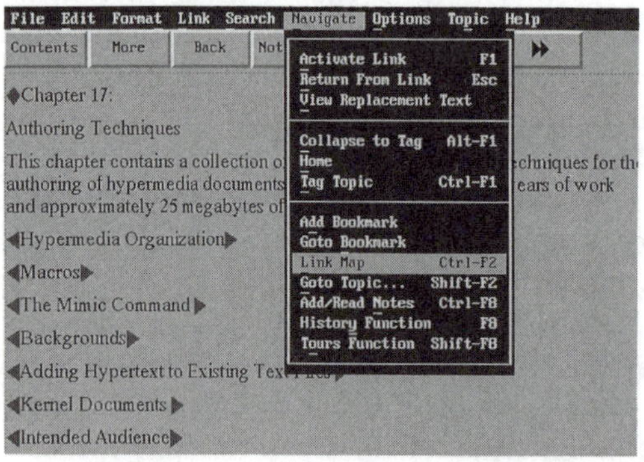

Figure 4–7
Screen shot highlighting the selection of navigational tools that may be supported by a program. (Software courtesy of Ntergaid, Inc.; HyperWriter!)

Another essential feature of HyperWriter and other hypermedia software is a program's *navigational* tools, which help a reader navigate through a document. In HyperWriter, different-colored symbols indicate specific link types. Another tool is a *link map*, a graphical representation that allows a reader to move rapidly to specific points in the document.

These navigational aids, among others, enhance the user–software interface. This is a vital consideration for any hypermedia author, since the information is made more accessible to the reader. Ultimately, the communications process is enhanced.

Hypertext and hypermedia programs also support other functions, such as a built-in text editor and the capability to import text and graphics from other programs. It may also be possible to alter text styles and colors, and with respect to HyperWriter, different visual backgrounds can be incorporated in a document for aesthetic and user-interface purposes.

Applications and Implications.   Hypermedia and/or hypertext systems are used in many applications. Software companies have, for instance, created help systems to support their programs. By clicking on a key term in an on-screen menu, you can retrieve linked information and read about an operation or procedure without opening a manual. This process can save time and simplify the information-search process.

In another application, a hypermedia program can serve as an engine for an information kiosk. By clicking on a specific area of a map displayed on a screen, you can retrieve information about the corresponding geographical location. This can include bus schedules or even the street names for the region. This type of application is particularly effective when used with a touch-screen interface instead of a mouse.

Hypermedia and hypertext software can also be used for electronic publishing. In one example, a document that incorporated expert-system capabilities was created for the United Nations. This project, called REGIS, was designed to make information about aquaculture in Africa accessible to readers.[16]

Hypermedia systems are also conducive to browsing. When you choose the links you want to follow, learning can become more dynamic, since you control what you read and what other trails you may follow. Depending upon the system, it may also be possible to guide the reader with a more predefined path through the hypermedia document so important information is not bypassed.

But despite all of its advantages, a hypermedia system, like other organizational and communications tools, could be used ineffectively. One such problem area is poor document design.

As is the case with a video or film production, a hypermedia project should be planned. Each link, like each frame, should serve a specific purpose. A reader can become lost in a document if there are too many consecutive links, just as a viewer can become lost in a television program if the plot is too confusing. A prototype of the final project, like a storyboard for a production, can help point out potential problems.

Another factor is that paper documents are still preferable for certain tasks. If you are conducting research about a topic, you can quickly scan through and refer to several or more books and magazines scattered on your desk. Even though this process can be somewhat duplicated with a computer, you may be limited by various hardware and software constraints. The scanning process may also be slower and more cumbersome.

For example, a monitor could display multiple windows of information. But without using a mouse to scroll down the different pages of text, your initial and overall view may be limited to only one or two paragraphs per window. With books and magazines, entire pages are visible at any given time and can be quickly scanned.

CONCLUSION

Through the use of hypertext and hypermedia programs, we can now create electronic documents that are limited only by our imagination and hardware and software constraints. Hypermedia programs can be potent educational tools: hypermedia can transform a static book into a living document.

Instead of simply reading about Beethoven, you can read about Beethoven while potentially seeing his picture and hearing his music. In the art world, instead of viewing just one side of a statue, you may be able to view all sides through an animation or a video clip. The various hypermedia links would enable you to explore these and other subjects in more depth and variety.

Hypermedia technology, when combined with optical-disc systems, will also contribute to the growth of our paperless society, in which information is increasingly created, exchanged, and stored in an electronic form. A key element of this development is a society no longer dependent on paper for writing and storing memos, checks, and documents of all kinds.

The concept of a paperless society must be examined from a broad perspective. Optical discs play a role in this overall system, as do the information services examined in Chapter 10. The development of high-speed communications lines that could handle large volumes of information, as well as the emergence of personal media as described in Chapter 8, are other contributing factors.

This switch to a paperless society could have a number of benefits, including the following:

1. We would witness the creation of an enhanced communications system. Information could be read, stored, and easily retrieved with the accuracy and control afforded by a computer.

2. We would be able to conserve physical space because information would no longer be saved on bulky paper.

3. Electronic publications—optical discs, or magazines published over communications lines—could be easily revised. This is an important factor for industries whose technology base can rapidly shift, necessitating continual updating of repair and operational manuals to keep pace with these changes.

The advent of cost-effective CD-R systems also opens up electronic publishing to a broader user base. For individuals or compa-

nies with mass-storage demands but not the mass-distribution requirement normally associated with CD-ROMs, a CD-R system may prove useful. You could create the disc yourself and use it with conventional CD-ROM players.[17]

4. Environmentally, fewer trees would be cut down for paper, and less waste material would be generated to further clog our landfills.

5. Information in an electronic form is very elastic. It can be stored diversely and organized and communicated through hypermedia and other retrieval mechanisms. Different types of information, such as video clips, graphics files, and digitized voice, can also work together, as in a multimedia presentation. This electronic document could subsequently be distributed by magnetic or optical media or relayed over a communications line.

6. A PC's processing power can be tapped to conduct searches for specific data. In one application, instead of looking through hundreds of pages of information to find a quote in a Sherlock Holmes novel, you can use a PC to conduct a search. Load in the appropriate CD-ROM, type in a keyword or words, and the PC can find the line. Depending on the interface, you may find not only the quote but also valuable complementary information.

7. As previously described, paper documents may be preferable for certain research tasks, such as quickly scanning though stacks of books and magazines. But the proper hardware and software setup could support an enhanced PC-based operation.

You could use the PC to find a specific term, click on a word, and bring the document up in another window. Or you could search through an on-screen table of contents, bring up a passage in one window, and quickly retrieve a graphic in another window.[18] You may also be able to create your own electronic notes about the information you are viewing.

8. Electronic book readers, pioneered by Sony's Data Discman, make electronic information portable. You could carry a reader with you and use it for reading and retrieving information from dictionaries and encyclopedias,

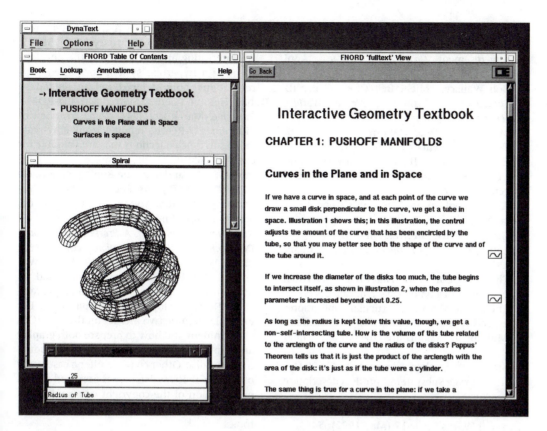

Figure 4–8
A sophisticated electronic book. Multiple views and hypertext links may be supported, including in this case a link to a program that illustrates the theorem mentioned in the text. It may also be possible to create your own electronic notes. (Courtesy of Electronic Book Technologies, Inc.)

travel guides, Shakespeare, and even movie compilations. Later models, including a portable CD-I player, extend these capabilities.

In sum, the growth of the paperless society does have positive benefits. There are, however, some potential pitfalls.

As examined in this chapter, paper documents still have some advantages. Newspapers, magazines, books, and other paper products are also cost-effective, readily mass-produced, and can be read in places where computer use is not feasible.

People are used to working with paper documents, and electronic information systems, unlike paper, mandate the use of an electronic device.

Another potential problem with the electronic environment is the standards issue. The very nature of electronic publishing calls for a series of machine-dependent operations to store, retrieve, and exchange information. It is crucial that industry-wide standards are adopted to ensure that a disc, for example, recorded on one manufacturer's equipment can be read on another device. If not, a customer with a stack of discs may not be able to retrieve the stored data if a manufacturer of a proprietary system goes out of business.

The growth of a paperless society also has privacy implications. The Lotus CD-ROM, previously mentioned, is only the tip of the proverbial iceberg (additional implications are covered in Chapter 10). Government regulation also plays a role in this area.

Finally, if information is reduced to an electronically mediated form, will everyone be able to gain access to it? If not, what are the potential ethical, economic, social, and political implications? Also, what is a realistic timetable for the technology to trickle down to all levels of society? If only a few can afford to use these new information systems, what are the implications for a democratic society?

REFERENCES/NOTES

1. Paul Freiberger, "CD Rot," *MPC World*, June/ July 1992, 34.
2. As of early 1992, the price was $20 for the

first disc and only $6 per additional disc in a series.

3. Advertisement, "CorelDraw," *NewMedia*, April 1992, back cover.

4. Bob Wallace, "Mich. Bell Offers Caller ID Service with Call Blocking," *Network World* 9 (March 2, 1992): 19.

5. Mary Martin, "Expectations on Ice," *Network World* 9 (September 7, 1992): 44.

6. Henry W. Jones, III, Esq., "Copyrights and CD-ROM," *CD-ROM Review*, November/December 1987, 64.

7. A more sophisticated "business" model was later released.

8. CD-Interactive Information Bureau, "CD-I Launches Titles," *CD-Interactive News*, January 1992, 4.

9. The Tandy Corporation later released its own version of this type of system, called VIS.

10. The CD-ROM XA has particular applications, especially in a multimedia environment. Multimedia systems are covered in Chapter 8.

11. Storage Technology Corporation, "7600 Optical Storage Subsystem," product description brochure.

12. Andy Reinhardt, "Floptical Arrives at Last, and It Works," *Byte* 17 (May 1992): 54.

13. Greg Harvey, *Understanding HyperCard* (San Francisco: Sybex, 1989), 3.

14. Ntergaid, Inc., "HyperWriter," flyer.

15. Ntergaid, Inc., *HyperWriter! User's Guide and Reference Manual* (Fairfield, Conn.: Ntergaid, Inc., 1991), 5. Please note: Topics are also referred to as *cards* or *nodes*.

16. Knowledge Garden, "New from Knowledge Inc.," information release.

17. Jon Udell, "Start the Presses," *Byte* 18 (February 1993): 132. Please note: The article provides a detailed look at CD-ROM publishing and the CD-R. Includes sidebars that cover topics ranging from optical-disc standards to CD-ROM drives.

18. Louis R. Reynolds and Steven J. Derose, "Electronic Books," *Byte* 17 (June 1992): 268.

ADDITIONAL READINGS

Evans, Ron. "Expert Systems and Hypercard." *Byte* 15 (January 1990): 317–24. Examines the use of HyperCard to create expert systems. Expert systems, which can provide a user with an in-house expert in the form of a computer program in various disciplines, are discussed in Chapter 3.

Foskett, William H. "Reg-In-A-Box: A Hypertext Solution." *AI Expert* 5 (February 1990): 38–45. An interesting article that traces the development of a hypertext system concerned with regulations about underground storage tanks.

Kobler, Helmut. "Getting Started with CD-ROM." *Publish* 6 (March 1991): 44–49. An overview of different CD-ROMs, ranging from gray-scale images to type collections that are valuable for desktop-publishing applications.

Lambert, Steve, and Suzanne Ropiequet, eds. *CD ROM: The New Papyrus*. Redmond, Wash.: Microsoft Press, 1986. And, Ropiequet, Suzanne, ed., with John Einberger and Bill Zoellick. *CD ROM Optical Publishing: A Practical Approach to Developing CD ROM Applications*, vol. 2. Redmond, Wash.: Microsoft Press, 1987. This two-volume CD-ROM book series, published by Microsoft Press, provides a comprehensive look at CD-ROM development and applications. For example, volume two focuses on the CD-ROM environment and how to prepare both graphic and alphanumeric information for reproduction on a disc. Other chapters offer a case study of how a CD-ROM database is prepared and an examination of the copyright issues pertinent to electronic publishing, among a variety of topics.

Nelson, Theodor. "Managing Immense Storage." *Byte* 13 (January 1988): 225–38. A description of the storage engine of the Xanadu project, a model for a new information-storage and -management system that could accommodate a range of information and could potentially create a powerful and global hypertext working environment.

Ravich, Leonard E. "Optical Storage Becomes Multifaceted." *Laser Focus World* 25 (March 1989): 115–22. And, Harvey, David A. "Optical Storage Primer." *Byte* 15 (Fall 1990; IBM Special Edition): 121–30. An examination of different optical-disc systems, including erasable configurations.

Waring, Becky, and Alexander Rosenberg. "New CD-ROM Hardware Swells the Consumer Market." *NewMedia*, May 1992, 12–15. A look at CD-ROM systems, including CD-I and CDTV configurations. The article contains sidebars on Kodak's Photo CD and the role of game manufacturers, such as Nintendo, in this market.

GLOSSARY

Compact disc (CD): In one form, a prerecorded digital optical disc that stores music. The CD player uses a laser to read the information stored on a disc.

Compact disc-interactive (CD-I): As initially introduced, a nonerasable, interactive, prerecorded digital optical-disc system designed for consumers.

Compact disc-read only memory (CD-ROM): A nonerasable, prerecorded digital optical-disc system that stores data. CD-ROM applications range from the distribution of computer software to electronic publishing. Recordable systems (the CD-R) also exist.

Erasable optical discs: A class of optical disc that acts much like conventional computer storage devices (for example, hard drives): data can be stored and erased. Different techniques can be employed for this operation.

Hypertext: A nonlinear system for information storage, management, and retrieval whereby links between associative information can be created and subsequently activated. The concepts behind a hypertext system have also been extended to include pictures and sounds to create a hypermedia environment.

Optical disc: The umbrella term for optical storage systems.

Paperless society: A society in which information is increasingly created, stored, and exchanged in an electronic form.

Photo CD: An optical disc that can store pictures. Both professional and consumer applications have been supported.

Videodisc: The pioneer, so to speak, of nonerasable, prerecorded optical-disc systems. Videodiscs are typically used for interactive training applications and to distribute movies (to consumers).

Write once, read many (WORM) optical disc: A WORM disc is a nonerasable but user-recordable digital optical-storage system.

5 | Fiber-Optic Technology

As outlined in Chapter 4, optical media that employ light as an element of the information-storage-and-retrieval process have been widely adopted. A similar development has taken place in the general communications industry. Fiber-optic systems, which use light to relay information, have been employed in various applications, carrying information ranging from a person's voice to the pictures produced by videocameras.

The concept of harnessing light as a modern communications tool stretches back to the late nineteenth century. Alexander Graham Bell, the inventor of the telephone, patented an invention in 1880 that used light as a means to transmit sound.[1] Bell's invention, the photophone, employed sunlight and a special light-sensitive device in the receiver to relay and subsequently reproduce a human voice.

It wasn't until the twentieth century, though, that the idea for such a tool could be transformed into a practical communications system. Two developments brought about this advance: the perfection of the laser and the manufacturing of hair-thin glass lines called fiber-optic (FO) lines. When combined, they helped create a *lightwave communications system*, a system in which modulated beams of light are used to carry or transmit information.

This chapter consequently focuses on FO technology and applications. The latter include the creation of sophisticated cable operations and the possible development of an integrated information and entertainment system. New services that could be carried on FO lines will also be examined. The chapter concludes with a discussion of holography, another technology that can employ light for numerous applications.

FIBER-OPTIC SYSTEMS

Light-Emitting Diodes and Laser Diodes

The integration of FO technology in the contemporary communications system has led to the creation of high-capacity communications lines. A transmission is conducted with optical energy, beams of light, produced by a special transmitter equipped with either a *light-emitting diode* (LED) or a *laser diode* (LD). The light is then confined to and carried by a highly pure glass fiber.

At this time, both LEDs and LDs are used in different communications configurations. An LED is less expensive and lasts longer than an LD, and it is generally well suited for relays of low-volume data over short distances. Like an LED, an LD is a semiconductor, but in the form of a laser on a chip. It is a small, powerful, and rugged semiconductor laser that is well suited for high-volume as well as medium- and long-distance relays. Other members of the LD family are also key elements of the optical storage systems described in Chapter 4.

The Fiber-Optic Transmission

In an FO transmission, a beam of light, an optical signal, serves as the information-carrying vehicle. Both analog and digital transmissions are supported, and in the case of a digital operation, the optical signal can be generated as individual pulses in an "on" and "off" pattern.

Next, the light is launched or fed into the fiber, the communications line for this system. The fiber itself is composed of two lay-

Figure 5–1
How a fiber-optic system works. The electrical signal is converted into an optical signal and back, following the relay. (Courtesy of Corning Inc.)

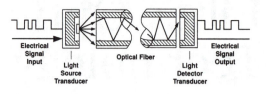

ers, the cladding and the core. Due to their different physical properties, light can travel down the fiber by a process called total internal reflection. In essence, the light travels through the fiber via a series of reflections that take place where the cladding and core meet, the cladding/core interface. When the light reaches the end of the line it is picked up by a light-sensitive receiver, and after a series of steps, the original signal is reproduced.

To sum up, in an FO system, an electrical signal, such as a videocamera's output, is converted into an optical signal. It is subsequently transmitted down the line and converted back into an electrical signal by the receiver.

Finally, the fiber, which may be made out of plastic in short-distance operations, is covered by a protective layer or jacket. This layer insulates the fiber from sharp objects and other hazards, and it can range from a light-protective coating to an armored surface de-

signed for military operations. The now-protected fiber is called an FO cable, and it consists of one or more fiber strands within the single cable enclosure.

Advantages. An FO line has a number of distinct advantages as a communications channel. In comparison with the radio frequency portion of the electromagnetic spectrum tapped by radio and television facilities, light—the information-carrying vehicle in an FO system—can accommodate an enormous volume of information because of light's higher frequency range and bandwidth capacity. Transmissions are in the gigabit-plus range (billions of bits per second), and in fact a single 0.75-inch fiber cable can replace 20 conventional 3.5-inch coaxial cables.[2]

FO lines are immune to electromagnetic and radio interference. Since light is used to convey the information, adjacent communications lines cannot adversely affect the transmission. Similarly, a system can be installed in a potentially explosive environment where gas fumes may build up over time.

An FO line also offers a higher degree of data security compared to copper-wire and cable systems. The nature of a fiber makes it extremely difficult to tap; if the fiber was ever tapped, the signal disturbance would be detected. Moreover, unlike other communications systems, FO lines do not radiate: a signal cannot be picked up by highly sensitive instruments unless the line is physically tapped.[3]

Also, information can be relayed a great distance without repeaters. A new generation of LDs, sensitive receivers, and fibers that match the characteristics of an LD's output have made it possible to relay signals more than 100 kilometers without repeaters. In practical terms, if fewer repeaters are used to create a system, the line will cost less to build and maintain.

An FO line is also a valuable asset for applications where space is at a premium, such as a building's duct space for carrying cable. Since fiber is comparatively narrow, it can usually fit in a space that may preclude the use of conventional cable.

Figure 5–2
An advantage of a fiber-optic line is its great information-carrying capacity. One strand of single-mode fiber can carry 8,000 simultaneous long-distance telephone conversations. New developments in this field, led by companies such as Corning, promise to continue to revolutionize the way we use and relay information. (Courtesy of Corning Inc.)

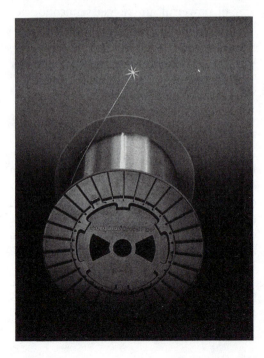

Disadvantages. Some factors do, however, limit an FO line's effectiveness as an information-carrying vehicle. As with other communications systems, there may be a loss of signal strength, which in this case may be due to physical and material properties and impurities.[4]

In another example, dispersion can affect the volume of information a line can accommodate in a given time period, that is, its channel capacity. In one form of dispersion, the paths or modes by which the light travels down the fiber can cause pulses to spread out or smear.[5] Thus, a part of one pulse may run into the tail end of another pulse. This collision will adversely affect the transmission since the information will not be properly received, and ultimately it will have an impact on a line's information-carrying capacity.

This factor can be addressed, in part, by using single-mode fiber. In contrast with multimode fiber, with which dispersion can be a chronic problem, the single-mode fiber is constructed with a much narrower core; the light essentially travels straight down the fiber in a single path. A single-mode fiber can accommodate a high information rate and has been the backbone of the telephone industry and other high-capacity long-distance systems.

Fiber is harder to splice than conventional lines, and the fiber ends must be accurately mated to ensure a clean transmission. But special connectors, among other devices and techniques, have facilitated this process.

Another disadvantage of fiber is the current state of the communications industry. Despite the inroads made by FO technology, the overall system is still dominated, in certain settings, by a copper standard. This includes the local telephone industry.

FO components can also be more expensive; a cable subscriber, for example, may pay more for a hookup using fiber rather than coaxial cable. The actual FO line may likewise cost more in specific applications, even though this price differential has diminished.

Thus, FO technology is analogous to digital technology in that it has been integrated in the current communications structure. The prices for LDs and the components that compose an FO system are also dropping, and the growing volume of information our communications systems must handle will mandate the construction of such high-capacity channels. There may also be hybrid systems, as described later in this chapter.

Applications

An FO line is an attractive transmission medium for the video-production industry. In one case, its lightweight design and transmission characteristics could make it a valuable in-field production tool. Several hundred feet of line, for instance, may weigh only several pounds, and a camera connected to the FO line could be used at a far greater distance from a remote van than a configuration with conventional cable.[6]

Fiber's information capacity also makes it an ideal candidate for an all-digital television studio. The digitization of broadcast-quality signals can generate millions of bits per second. An FO system would be capable of handling this great quantity of information. The line would also fit in a studio's duct space, which is usually crammed with other cables, and its freedom from electromagnetic and radio frequency interference would be valuable assets in this environment.

In addition to its role in the television field, FO systems have been adopted by a wide range of industries. In one application, an FO link makes it possible to create very efficient and high-capacity LANs. Businesses, hospitals, schools, and other organizations are likewise designing and implementing their own FO lines that can accommodate computer data as well as voice and video information.

The telephone industry has a vital interest in this technology. Long-distance carriers have developed extensive FO networks, and eventually fiber may even be extended to our homes.

Fiber may also provide us with new high-speed data highways. As we generate not only more information, but more complex information with each passing year, the communications infrastructure may be hard pressed to handle this data flow.

But FO technology has played and will continue to play a key role in solving this problem. In one example, optical highways may tie the country in a communications net to facilitate the exchange of all types of information between industry, researchers, and educators. As described in a later section of this chapter, FO lines may also be the backbone of a new type of service that could provide consumers with an array of new television and information options.

Underwater Lines. Beyond these landline configurations, AT&T headed an international consortium that developed the first transatlantic undersea FO link between the United States and Europe. The system, called TAT-8, is more than 3,000 nautical miles in

length and was the first transoceanic FO cable.

TAT-8, which serves as a model for this type of communications tool, is a sophisticated configuration made possible by the selection of a high-capacity FO line as the transmission medium. TAT-8 was designed to accommodate a variety of information, and when inaugurated, the system had an estimated lifetime in excess of 20 years.[7]

Even though fiber had already been used in long- and short-distance land and undersea operations, respectively, TAT-8 was the first of a new class of undersea cables. Its installation was preceded by extensive deepwater experiments and trials conducted in the early 1980s to demonstrate the project's feasibility. Once completed, the findings confirmed the expectations of the system's designers. The FO cable and repeater used in the experiments relayed the information well within acceptable error rates and transmission losses even while subject to the ocean's extreme environmental conditions, such as the cold temperature and tremendous water pressure. The cable also survived the physical strain imposed by the laying and recovery operations.

Like other new applications, though, FO lines have had some problems. Certain TAT-8 users, for instance, experienced some outages due to cable damage caused by fishing trawlers.[8]

FO Lines and Satellites. At first glance, an undersea cable and, in fact, long-distance land lines, may appear to be obsolete communications channels in light of the established satellite networks. Yet an FO system has a wide channel capacity, an extended lifetime, and can be cost-effective for long-distance applications.

An FO line also has certain advantages over a satellite. A satellite transmission may be susceptible to atmospheric conditions; also, there's a slight time delay inherent in satellite traffic, which could be disruptive in certain types of data relays.[9] An FO link, in contrast, is not subject to either constraint.

Fiber is also a more secure conduit for sensitive material. A satellite transmission's broad reception area makes it possible for

Figure 5–3
The original TAT-8 configuration has subsequently blossomed into a number of planned fiber-optic systems. (Courtesy of KMI Corp., Newport, R.I.)

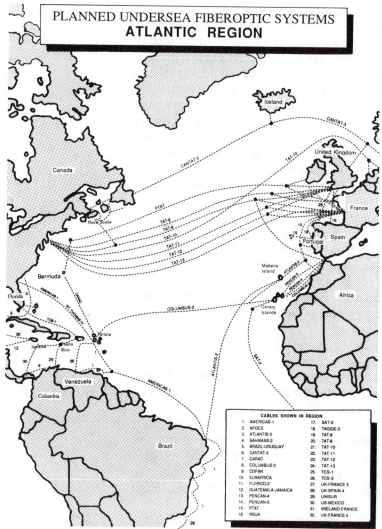

© 1992 KESSLER MARKETING INTELLIGENCE

unauthorized individuals to receive its signal. While it's true the signal could be encrypted and coded for protection, it's equally true that another person could decode the transmission once it has been intercepted.

Finally, it appeared, at one point, that satellites would completely dominate all long-distance communications traffic. This scenario has been radically altered, however, in view of the development of the new and sophisticated FO lines. It now appears that fiber and satellite systems will actually supplement and complement each other on both national and international levels.

A satellite, for its part, can support a flexible point-to-multipoint operation that can readily accommodate additional receiving sites. This is not the case with an FO system, in which a special line may have to be laid to reach the new location.

Yet fiber is ideally suited for intracity relays and specific types of point-to-point communication. It has been used for television network news relays, by the telephone industry, and in teleconferencing applications. In one ambitious plan, it may constitute an underwater link between Japan and the United Kingdom that may cover some 15,000 miles.[10] Thus, each field has its own particular strengths, and both systems have been and will continue to be used to enhance our communications capabilities.

Other Applications. Besides these mainstream activities, FO technology has been adopted for more esoteric applications. In the medical field, for instance, an FO line can be used in certain types of laser surgery. The fiber can act as a vehicle to transport the intense beam of light to the operating site. In a related application, fibers serve as a visual inspection tool in the form of a *fiberscope*. This device "consists of two bundles of optical fibers. One, the illuminating bundle, carries light to the tissues, and the other, the imaging bundle, transmits the image to the observer."[11] This device enables doctors to peer inside the human body.

Scientists are also experimenting with FO sensors to monitor the physical condition of composite materials. A relatively new industry has developed that employs nonmetallic

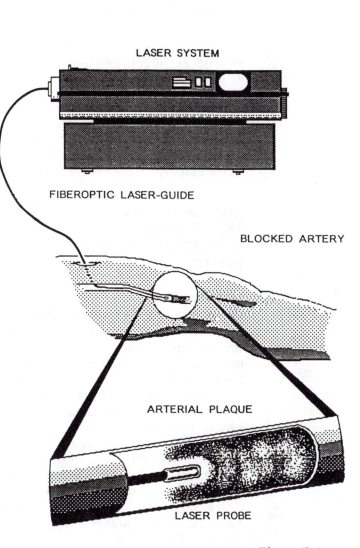

LASER ANGIOPLASTY

LASER SYSTEM

FIBEROPTIC LASER-GUIDE

BLOCKED ARTERY

ARTERIAL PLAQUE

LASER PROBE

Figure 5–4
A medical application of fiber-optic technology. (Courtesy of KMI Corp., Newport, R.I.)

composite materials to create items ranging from aircraft parts to, potentially, a space station. A fiber can literally be embedded in such a structure, and based upon certain test procedures, it would be possible to determine if the structure was subject to damaging and undue stress.[12]

While such a fiber may not be relaying computer data, it is, nevertheless, providing observers with vital information. As discussed in the beginning of the book, this is one of the key features of the communication revolution. Information will take new forms and shapes, and in this particular case, the data about a structure's physical condition does, in fact, convey an intelligence.

All the applications described thus far only touch on the different roles FO technology can play in society. In fact, it has even been explored by the aerospace industry for fly-by-light configurations where conventional, heavier cable would be replaced by fiber for command-and-control operations. The fiber would save weight, and its freedom from electromagnetic interference would lend itself to this type of environment.[13] The automobile industry has also explored this option.

THE CABLE AND TELEPHONE INDUSTRIES AND YOUR HOME

Introduction

Fiber may serve as the backbone of enhanced entertainment and information systems that could feature television programming, stock market quotes, and other services. These offerings could possibly be supported by cable or telephone companies (telcos) and, in fact, may even be distributed via coaxial cable or a hybrid configuration.

The key element in this system is providing individuals with more control over their programming choices, be it a television show or a new optional service. Two terms have been associated with this development, *video on demand* (VOD) and *video dial tone* (VDT), also spelled *video dialtone*.[14]

In brief, VOD could be viewed as an enhanced *pay-per-view* (PPV) option. With PPV, you may pay a cable company a set fee to view a movie or a special event, such as a concert, typically on a designated PPV channel(s).

VOD would expand this capability by supporting a more diverse and tailored programming pool. At its highest level of sophistication, VOD could function much like a video rental store. You select the programming you want to see when you want to see it.

With VOD, for instance, an individual would have multiple programming options. These could include self-help and exercise videotapes, movies, and a library of older television programs.

In one configuration, which can be viewed as enhanced PPV, more movies would be available on shorter notice than with current operations. In a higher-level setup, movie libraries could be made available and a service could support various interactive features. This may include a pause button that would function much like a pause mechanism on a conventional VCR.[15]

VDT, on the other hand, is designed to allow telcos to compete in the video marketplace. Options could include VOD programming, some of the information services described in Chapter 10, and as will be discussed, other communications offerings.

Since the 1980s, the broadcasting and cable industries haven't relished the idea of the telcos competing in their own backyards. At that time the telcos were blocked from participating in these and related fields.

For example, as part of the ongoing evaluation of the consent decree that led to the divestiture of AT&T, Judge Harold Greene, the architect of this decision, barred the seven regional holding companies from offering various information services. The companies were created after the breakup of AT&T, and the restrictions were initially adopted to prevent the telephone industry from dominating the new, emerging information infrastructure.

But political pressures and other factors led to a relaxation of the original decision in the late 1980s and early 1990s. This opened the door for the telcos to support information services, such as electronic Yellow Pages, and potentially to enter the video market.[16]

The next major step was VDT. In 1992 the FCC modified its rules to allow telephone companies to compete in the video arena. One provision called for the support of a "basic platform that will deliver video programming and potentially other services to end-users" on a common carrier, that is, a nondiscriminatory basis.[17] In essence, the telcos could now carry video programming, with certain restrictions. Another recommendation called for the repeal of the telco-cable cross-ownership rules.[18]

One of the goals behind VDT was the enhancement of the current communications infrastructure. It was hoped that VDT would provide the impetus for the development of an advanced system, which in turn could offer consumers new services that could be transparently and easily accessed.[19]

Finally, both VOD and VDT reflect one of the apparent fallouts of the new communications technologies. As described in other chapters, rather than serving the mass audience, individuals with their individual needs would now be satisfied. In one sense, the communication revolution could be viewed as a personal revolution, and this integrated entertainment and information utility could play an important role in the new communications structure.

Applications and Implications

The idea of a comprehensive entertainment and information utility did not begin with VOD and VDT. The Qube cable-television service, launched in Columbus, Ohio, in the 1970s, let subscribers tailor their cable-television service to match their particular viewing requirements.

Its most unique characteristic was an interactive capability. After a speech delivered by then-President Jimmy Carter, for instance, Qube subscribers participated in an electronic survey by responding to a series of questions that appeared on the television screen. The subscribers registered their answers via the keypads, and Qube's computer tabulated the results.[20]

Similar viewer-response programs were produced in other subject areas, and the system had the capability to support a PPV option, information services, and electronic transactions. The latter could include shopping and banking at home.

Even though Qube later abandoned its ambitious plans, this same concept, of an integrated entertainment and information utility, could become a reality through new cable and/or telephone links. For example, experimental VOD systems have been developed, and during the early 1990s, the cable industry rushed in to support different levels of this application.

Other key issues, which may play a role in shaping this potential service, include the following:

1. True or real-time VOD would demand enhanced and *addressable* converter boxes that would be more sophisticated than the typical unit in current use. In contrast to a service where all the converters may be the same, a more advanced box would be individually identified by an address or code. Under this system, specific programming, such as movies, would be delivered to only those boxes with the correct addresses, per subscriber requests.

In fact, the next generation boxes could resemble computers in regard to their sophistication and user interface. Various computer companies have also contributed to the development of systems that could support VOD, among other operations.

2. A cable system must have the channel capacity to support additional services. In one case, it's possible to construct fiber/coaxial hybrid systems.

Fiber could, for example, be used in a trunk line and coaxial cable used in other parts of the system. In another configuration, fiber could be directed to a "curbside terminal shared by four or more users."[21] Standard cable would then be used for the actual links to the homes.

The hybrid systems would benefit, in part, from the transmission advantages and channel capacity afforded by fiber and the cost-effectiveness of retaining the current technology base for individual subscribers. In one implementation, a hybrid system has initially provided subscribers with 150 channels.[22]

3. Even though it appears that the trend is for the eventual implementation of all-fiber systems, the timetable is still unknown. The cost for a full switchover would be very expensive, and as described, fiber/coaxial hybrids and/or data compression may prove suitable, at least in the interim.

Digital compression could increase the number of channels a cable system currently supports. Consequently, instead of building a new FO plant, a coaxial plant may be able to handle more-advanced services.

For telephone companies, digital compression may make it possible to use existing telephone lines to deliver video programming. This could be the first step in the development of an all-fiber system and would provide the telcos with immediate access to most homes in the United States.[23]

4. If fully implemented, VOD and VDT operations could fulfill the promise of the original Qube system, with additional features to boot. Besides television programming, the interactive information-based services described in Chapter 10 could be supported.

Other potential services also exist. These range from delivering multimedia presentations (outlined in Chapter 8) to supporting videoconferences (covered in Chapter 11). Basically, if a system has excess capacity, it could be used in numerous applications.

New and emerging technologies may also support this field. In one example, the high-definition television systems described in Chapter 9 promise to deliver high-quality pictures and sounds to our homes. When combined with VOD, your living room could be converted into a movie theater.

5. Despite its promise, there are some potential stumbling blocks in the full implementation of VOD services, including wide consumer acceptance.

While VOD may be quickly adopted by specific demographic groups, the question about consumer demand on a broad scale is still unanswered. How much money are people willing to spend for VOD services? Can these services be sustained by targeting select subscriber groups via narrowcasting? How many years will it take for the necessary technologies to reach smaller cable markets? Are the changes inevitable, as is the case with most industries where a physical plant is modernized over time, or will some cable companies balk at the additional expense?

6. Another factor is competing services. Satellites could potentially deliver a wide variety of programming to subscribers, as discussed in the next chapter, and even television stations could emerge as potential competitors for multiple programming choices. The latter topic is outlined in Chapter 9.

There may also be more cooperative ventures, and possibly mergers, between different industries. In one operation, New York Telephone and Liberty Cable Television, a wireless operator in Manhattan, developed a plan to deliver programming.[24] Liberty would use New York Telephone's FO lines to deliver enhanced services to two apartment buildings.

7. The telephone industry's overall role in this endeavor is still not clear, as of this writing. VDT did open the video-programming door. Yet the telcos still pressed for the "freedom not only to be common carrier purveyors of video, but also to be full-blown cable operators with the *right to own and package programming* [my emphasis]."[25] Consequently, they asked for a further relaxation of the legal restrictions that existed at the time.

The NTIA had essentially supported this approach, with built-in safeguards to help prevent potential abuses and the development of a noncompetitive market. Ultimately, the goal was to promote diversity and the enhancement of the current infrastructure by allowing the telcos to participate fully in these new ventures: by expanding the telcos' operational limits, they would have an incentive to build new, high-capacity FO communications channels.[26]

8. The cable industry is in an historically stronger position than the telcos to provide programming. This advantage, and the capability to use new techniques to possibly increase a cable system's channel capacity without a major overhaul of the entire physical plant, may prove to be formidable obstacles for the telcos to overcome.

Yet if compression proves successful, the telcos, for their part, may have an even wider market penetration via their telephone lines, even though a fiber-based system may have to be built to fully compete with the cable companies. The telcos also have the necessary financial resources for launching new operations.

Conclusion

The development of FO systems promises to alter the way we communicate. On one front, FO systems have already created sophisticated national and international telephone links. On another front, they promise to change the way we view television and, potentially, the way we receive and use information.

The latter include, as indicated, the information services depicted in Chapter 10. In this configuration, the same system could handle telephone calls and potentially relay, over high-speed lines, information ranging from stock price quotes to support material for college courses you can take at home.

Unlike the current telephone-delivered operations, this material could encompass text and video, and the system may even carry interactive electronic meetings, the teleconferences discussed in Chapter 11. While separate communications lines could accommodate this range of information, an integrated service could enhance the process and help make the concept behind a Qube-type system an operational reality.

But widespread consumer acceptance is still an unknown factor, as is the telcos' role. There is also a question about the time frame. In essence, when will VOD and VDT technology trickle down to the smaller markets, if at all?

The growth of FO lines, as well as VOD and VDT services, can also be viewed as evolutionary rather than revolutionary developments. For example, fiber was initially used in the telephone industry for long-distance lines, and now it may find its way to the local plant.

Consequently, the infrastructure is modernized, but the current system is not immediately abandoned, especially in light of its new potential. This includes the support of additional applications through compression techniques.

Eventually, this enhancement will enable the United States, or another country with a similar program, to better compete in the world market. It can also help bring the information age right to our doorsteps.

HOLOGRAPHY

Holography is one of the more interesting technologies discussed thus far. The actual product, a *hologram*, which can be considered a record of the optical information that composes a scene, has been used for everything from advertising to security.

Holographic techniques were initially discovered in the 1940s by Dr. Dennis Gabor, a research engineer, who won a Nobel prize in 1971 for his pioneering work as the father and creator of holography. But despite its earlier discovery date, it was not until the 1960s that holography could fully blossom, due to the development of a powerful and suitable light source to create holograms, the laser.

Even though it is beyond the scope of this overview to explore the physical properties of holography, we can still present a working definition. A hologram, a product of holography, is capable of storing information about a three-dimensional object. Furthermore, in contrast with a standard photograph, which

records light intensity . . . the hologram has the added information of phase . . . to show depth. When a person looks at a tree . . . he is using his eyes to capture light bouncing off the object and then processing the information to give it meaning. A hologram is just a convenient way to re-create the same light waves that would come from an object if it were actually there.[27]

This capability to show depth can be startling. Objects are quite lifelike, and they may appear to "jump out" of the scene. In addition, unlike a standard photograph, which records the scene from a single perspective, a hologram breaks these boundaries.

Finally, many holograms are created through a special type of film and are chemically processed. But newer techniques have been introduced to simplify and speed up this task, and holograms have been electronically produced on a reusable recording medium. This operation opened up the field to a broader user base.

Applications

An object, such as a statue, can be recorded as a cylindrically shaped hologram. As you walk around the hologram, you will see the recorded statue as you would the original object—from various views. This type of reproduction could be a valuable storage medium for preserving the characteristics of various three-dimensional art objects and for distributing holographic copies of an object for research purposes.

The advertising and security fields are also supported by this technology. Attention-grabbing ads can be produced, and small, inexpensive embossed holograms have been placed on credit cards to curtail the proliferation of counterfeit cards. It would be very difficult to duplicate this type of hologram and would require a high level of expertise as well as a fairly sophisticated physical plant.

Another application area is that of *holographic interferometry*. Unlike advertising and other artistic endeavors, a conventional three-dimensional view of an object is not created. Rather, in one variation of this process, a double exposure of the same object is made. During one exposure, the object is stationary and at rest. In the second exposure, the object is subjected to stress through the introduction of the physical forces it may experience in actual operation.

When processed, the hologram will depict a series of fringes or lines, which resemble the contour lines on a map, across the object. The fringes reveal the very small differences between the exposures, which in this case are the differences between the object while at rest and under stress.[28] The fringes, therefore, depict areas of the object that may be deformed by the operation, and the hologram serves as a visual map of the object's physical state or condition.

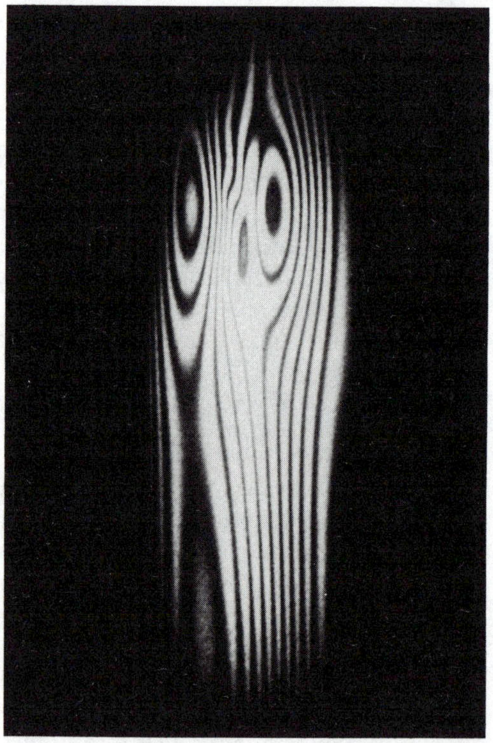

Holographic interferometry has been used for testing applications. In this operation, various products have been inspected for manufacturing defects, while metals and other materials have been subjected to stress to test for possible flaws.[29]

Holography has also provided us with another valuable tool, the *Holographic Optical Element* (HOE). A hologram, in this situation, functions as an optical element, such as a lens or mirror.[30] HOEs are very useful in certain situations, especially where a standard optical component may not fit or may be too heavy.

One application area has been the use of HOEs in head-up displays. While flying, an image of an instrument panel could be displayed before a pilot's eyes. Thus, a pilot would not have to move his or her eyes to view the instruments.

The holographic field may also support data-storage operations. Due to its unique properties, a holographic-based system could emerge as a highly effective storage system.[31]

The future may also witness the development of a 3-D television configuration. Experimental work in this area has already been conducted, and if combined with a CD-quality sound capability, the system would create a new generation of realistic computer games as well as entertainment and educational programs.

CONCLUSION

Holography, and especially FO configurations, play an important role in our communications system. As in other areas, though, developments promise to improve their capabilities. In the case of FO technology, this includes enhancements in the marriage between optical and electronic components (optoelectronics) and in the use of light alone.

In one example, optical amplifiers could

Figure 5–5

Corrosion monitored: an example of holographic interferometry. Holography can, in effect, let you visualize the interior of a pipe from the outside. The circular fringe pattern, due to a small pressure change, indicates weakened regions of the pipe wall. (Courtesy of the Newport Corp.)

improve an FO system's performance. In contrast to a current configuration, where the optical signal is converted into an electrical signal and back at a repeater site, the signal could remain in the optical domain.[32]

A similar development is taking place in the switching field, where an optical configuration would have an advantage over a nonoptical system.[33] Light may also be used to help produce newer, faster computers.

In sum, while these developments are exciting, we are only beginning to tap into light's potential as a communications tool. Its role in our information society cannot and should not be underestimated.

Finally, the next chapter explores another important communications system, our satellite network. Satellites and FO systems are complementary in many ways, and when viewed together, they are two of the driving forces behind the communication revolution.

REFERENCES/NOTES

1. Richard S. Shuford, "An Introduction to Fiber Optics," *Byte* 9 (December 1984): 121.
2. Arthur Parsons, "Why Light Pulses Are Replacing Electrical Pulses in Creating Higher-Speed Transmission Systems," *Communications News*, August 1987, 25.
3. "Optical Fiber Technology: Providing Solutions to Military Requirements," *Guidelines*, 1991, 6.
4. The latter factor, though, has been greatly reduced.
5. Ronald Ohlhaber and David Watson, "Fiber Optic Technology and Applications," *Electronic Imaging* 2 (August 1983): 29.
6. Richard Cerny, "Fiber-Optic Systems Improve Broadcast ENG/EFP Results," *Communications News*, April 1983, 60.
7. R. E. Wagner, S. M. Abbott, R. F. Gleason, et al., "Lightwave Undersea Cable System," reprint of a paper presented at the IEEE International Conference on Communications, June 13–17, 1982, Philadelphia, Penn., 7D.6.5.
8. Barton Crockett, "Problems Plaguing Undersea Fiber Links Raise Concern Among Users," *Network World* 20 (August 21, 1989): 5.
9. As described in the next chapter, a communications satellite is typically placed in an orbit some 22,000 miles above the Earth. The round-trip to and from the satellite results in the slight time delay.
10. "Presstime Bulletin; Agreement of World's Longest Undersea FO Cable," *Photonics Spectra* 26 (October 1992): 16.
11. Abraham Katzir, "Optical Fibers in Medicine," *Scientific American* (May 1989): 120. Please note: A fiberscope may also be incorporated in an endoscope, which provides physicians with remote access to the body regions under observation. The article presents an excellent overview of this field.
12. Richard Mack, "Fiber Sensors Provide Key for Monitoring Stresses in Composite Materials," *Laser Focus/Electro-Optics* 23 (May 1987): 122.
13. Luis Figueroa, C. S. Hong, Glen E. Miller, et al., "Photonics Technology for Aerospace Applications," *Photonics Spectra* 25 (July 1991): 117.
14. VDT, used in the context of computer systems, is an acronym for *video display terminal*. For this specific discussion, though, it represents *video dial tone*.
15. Richard L. Worsnop, "Pay-Per-View TV," *CQ Researcher* 1 (October 4, 1991): 743.
16. Steve Higgins, "NYNEX, Pacific Bell Tout Electronic Information Services," *PC Week* 9 (March 9, 1992): 67.
17. FCC, CC Docket 87–266, July 16, 1992, "Local Telephone Companies to Be Allowed to Offer Video Dialtone Services; Repeal of Statutory Telco-Cable Prohibition Recommended to Congress," downloaded from CompuServe, October 1992. Please note: There has been a call for a VDT-type option for a number of years.
18. Harry C. Martin, "Telcos Offer Video Services," *Broadcast Engineering* 34 (September 1992): 8.
19. FCC, CC Docket 87–266, July 16, 1992, "Local Telephone Companies."
20. Edward Meadows, "Why TV Sets Do More in Columbus, Ohio," *Fortune*, October 6, 1980, 67.
21. Larry Aiello, Jr., "Bring Fiber Home," *Guidelines*, 1992, 7.
22. "TW's Plan for Queens: 150 Channels, 40 of PPV," *Broadcasting*, March 11, 1991, 21. Please note: Besides replacing lines, other costs for an all-fiber operation would include the necessary optoelectronic components (that is, components used to convert from an optical to an electrical signal). Cable also has a very wide

channel capacity, especially when used for short distances. This factor helps make hybrid systems feasible.

23. "Telcos Eye Twisted Pair Video," *TV Technology* 10 (August 1992): 1.

24. Rich Brown, "New York Connects to Video Dialtone," *Broadcasting*, October 12, 1992, 38.

25. Harry A. Jessell, "FCC Calls for Telco TV," *Broadcasting*, July 20, 1992, 3.

26. National Telecommunications and Information Administration, "Telecommunications in the Age of Information," NTIA Special Publication 91–26, October 1991, 234.

27. Brad Sharpe, "Hologram Views," *Advanced Imaging* 2 (August 1987): 28.

28. W. Thomas Cathey, *Optical Information Processing and Holography* (New York: John Wiley and Sons, 1974), 324.

29. James D. Trolinger, "Outlook for Holography Strong as Applications Achieve Success," *Laser Focus/Electro-Optics* 22 (July 1986): 84.

30. Jose R. Magarinos and Daniel J. Coleman, "Holographic Mirrors," *Optical Engineering* 24 (September/October 1985): 769.

31. Tom Parish, "Crystal Clear Storage," *Byte* 15 (November 1990): 283.

32. Maureen Molloy, "Bell Labs Study Confirms Gains from Undersea Optics," *Network World* 8 (May 13, 1991): 31.

33. Paul R. Prucnal and Philippe A. Perrier, "Self-routing Photonic Switching with Optically Processed Control," *Optical Engineering* 29 (March 1990): 181.

ADDITIONAL READINGS

Amadesi, S., A. D'Altorio, and D. Paoletti. "Real-Time Holography for Microcrack Detection in Ancient Gold Paintings." *Optical Engineering* 22 (September/October 1983): 660–62. And, Carelli, Pasquale, Domenica Paoletti, and Giuseppe Schirripa Spagnolo. "Holographic Contouring Method: Application to Automatic Measurements of Surface Defects in Artwork." *Optical Engineering* 30 (September 1991): 1294–98. Holography and its uses as an inspection tool in the art/art-restoration worlds.

"Breakup of the Bell System." *Communications News*, September 1984, 98–99. And, Tanzillo, Kevin. "Flood Gates Open." *Communications News*. January 1989, 48–51, 65. The first article describes the divestiture of AT&T; the second

looks at the state of the industry five years after the fact.

Broadcasting. Broadcasting *magazine has chronicled the potential entry of the telcos in the video market. Sample articles include*

"Bell Atlantic Sees Television in Its Future." May 8, 1989, 30.

"Telco's Army Poised for Assault on TV Entry." October 3, 1988, 38–47. Please note: a profile of the top telcos is included in this special report.

"Fritts Warns Broadcasters: The Telcos Are Coming." September 12, 1988, 27.

"Advance Man for Telco Entry." July 4, 1988, 35–38.

Corning Incorporated, Opto-Electronics Group. *Just the Facts.* Corning, N.Y.: Corning, Inc., 1992. And, Karley, Brent A. "Fiber Optics." In Pohlmann, Ken C., ed., *Advanced Digital Audio.* Carmel, Ind.: SAMS, 1991. Both publications provide excellent overviews of FO technology and applications.

Geller, Henry. *Fiber Optics: An Opportunity for a New Policy?* Washington, D.C.: The Annenberg Washington Program, 1991. An in-depth report that examines FO systems, regulatory issues, telcos, and the "need to take into account First Amendment considerations," with respect to policy questions.

Hargadon, Tom. "Video Dial Tone: Let RBOCs Stick to Knitting." *NewMedia.* November 1992, 14. VDT and the potential applications, such as multimedia, that telcos could be pursuing.

Heath Company. *Heathkit Educational Systems-Fiber Optics.* Benton Harbor, Mich.: Heath Company, 1986. An excellent hands-on introduction to fiber-optic technology and systems. The product consists of a comprehensive educational manual that provides an overview of fiber-optic technology and applications. An accompanying set of hardware, including fiber-optic cable, can be used to conduct a series of experiments to enhance the learning process.

Hecht, Jeff. "Optical Computers." *High Technology* 7 (February 1987): 44–49.

"Optical Computer: Is Concept Becoming Reality?" *OE Reports* 75 (March 1990): 1–2. And, Silvernail, Lauren P. "Optical Computing: Does Its Promise Justify the Present Hype." *Photonics Spectra* 24 (September 1990): 127–29. A look at optical computing.

Hutchins, Don, and S. J. Campanella. "Is Fiber a Threat to Satellite-Based Nets?" *Network World* 4 (March 23, 1987): 26–27. A pro and con discussion of the fiber-optic vs. satellite systems debate.

Ih, Charles S. "A Holographic Process for Color Motion-Picture Preservation." *SMPTE Journal* 87 (December 1978): 832–34. An interesting look at using holography as a film-preservation system.

Jeong, Tung H., and Harry C. Knowles. *Holography Using a Helium-Neon Laser.* Bellmawr, N.J.: Metrologic Instrument Corp., 1987. A manual that explores general holographic principles and holographic production techniques. The manual is a companion to a holography kit manufactured by Metrologic. The kit included the necessary components, minus the laser, to produce a range of different holograms.

Katzir, Abraham. "Optical Fibers in Medical Applications." *Laser Focus/Electro Optics* 22 (May 1986): 94–110. A comprehensive look at the role and applications of fiber-optic technology in the medical field. The article cited in note 11 presents a more up-to-date examination of this field.

Moore, Emery L., and Ramon P. De Paula. "Inertial Sensing." *Advanced Imaging* 2 (November 1987): A48–A50. And, "Fiber Optics for Astronomy." *Sky and Telescope* 81 (December 1989): 569–70. Two nonmainstream FO applications.

Parker, R. J., and D. G. Jones. "Holography in an Industrial Environment." *Optical Engineering* 27 (January 1988): 55–66.

Rosenthal, David, and Rudy Garza. "Holographic NDT and the Real World." *Photonics Spectra* 21 (December 1987): 105–106. Industrial applications of holography.

Rosenthal, Steve. "Interactive TV: The Gold Rush Is On." *NewMedia.* December 1992, 27–29. And, Worsnop, Richard L. "Pay-Per-View TV." *CQ Researcher* 1 (October 4, 1991): 731–51. The articles cover different interactive television and PPV options, respectively.

GLOSSARY

Fiber-optic line: A highly pure, hair-thin glass fiber that is used as a conduit, a fiber-optic line, to relay a wide range of information (for example, voice and video). A fiber-optic line is a superior communications system in terms of its channel capacity and signal quality.

Hologram: A hologram is a record of the optical information that composes a scene. A hologram can be used for applications ranging from advertising to security functions.

Holographic Optical Element (HOE): An application in which a hologram functions as an optical element, such as a lens.

Laser diode (LD): One of the more versatile communications tools. The LD family supports products and applications ranging from CD players to fiber-optic systems.

Light-emitting diode (LED): One of the light sources for a fiber-optic system.

Single-mode fiber: An efficient fiber-optic communications line that can support high-speed communications relays.

Two-way cable: An interactive two-way cable system. Qube, in Columbus, Ohio, was an interactive two-way cable service that supported a variety of interactive functions.

Video dial tone (VDT): VDT is designed to enable telephone companies to compete in the video marketplace.

Video on demand (VOD): VOD can be viewed as an enhanced pay-per-view option. At one level, you could access and subsequently view a videotape of your choice from a library of features (for example, movies).

6

Satellites

Satellite communication has become a part of everyday life. We can now make an international telephone call as easily as a local call down the block. We also see international events, such as an election in England or a tennis match in France, with the same regularity as local political and sporting events.

This capability to exchange information on a global basis, be it a telephone call or a news story, is made possible through a powerful communications tool: the satellite. For those of us who grew up before the space age had dawned, satellite-based communication is the culmination of a dream. It stretches back to an era when the term *satellite* was only an idea conceived by a few inspired individuals. These pioneers included authors such as Arthur C. Clarke, who in 1945 fostered the idea of a worldwide satellite system. This idea has subsequently blossomed into a sophisticated satellite network that spans the globe.

The first generation of satellites was fairly primitive compared to contemporary spacecraft. These early satellites embodied *active* and *passive* designs. A passive satellite, such as the Echo I spacecraft launched in 1960, was not equipped with a two-way transmission system. Echo was a large aluminumized-mylar balloon that functioned as a reflector. After the satellite was placed in a low Earth orbit, signals relayed to Echo reflected or bounced off its surface and returned to Earth.

In contrast with the Echo series, the Telstar I active communications satellite, launched in 1962, carried receiving and transmitting equipment. It was an active participant in the reception-transmission process. As the satellite received a signal from a *ground* or *Earth station*, a communications complex that transmitted and/or received satellite signals, it relayed its own signal to Earth. Telstar also paved the way for today's communications spacecraft, since it created the world's first international satellite television link.

SATELLITE TECHNOLOGY

Satellite Fundamentals

Geostationary Orbits. During the span of years that separates Telstar I from today's satellites, there have been a number of improvements. Spacecraft such as Telstar and Echo were placed in low Earth orbits. In this position, a satellite traveled at such a great rate of speed that it was visible to an individual ground station for only a limited time each day. The satellite appeared from below the horizon, raced across the sky, and then disappeared below the opposite horizon.

Since the ground station was cut off from the now-invisible satellite, a station situated below the horizon had to be activated to maintain the communications link. Or it would have been necessary to launch a series of satellites to create a continuous satellite-based relay for any given Earth station. As one satellite disappeared, it would have been replaced by the next satellite rising on the horizon.

The latter type of satellite system would have entailed the development of a complex and cumbersome Earth- and space-based network. Fortunately, though, this problem was solved in 1963 and 1964 through the launching of the Syncom satellites. Rather than circling the Earth at a rapid rate of speed, the spacecraft appeared to be stationary, or fixed, in the sky. Today's communications satellites, for the most part, have followed suit and are now placed in what are called *geostationary* orbital positions, or *slots*.

Simply stated, a satellite in a geostationary orbital position appears to be fixed over one portion of the Earth. At an altitude of 22,300 miles above the Earth's equator, a satellite travels at the same speed at which the Earth rotates, and its motion is synchronized with the Earth's rotation. Even though the satellite is moving at an enormous rate of speed, it is stationary in the sky relative to an observer on the Earth.

The primary value of a satellite in this orbit is its ability to communicate with ground stations in its coverage area 24 hours a day. This orbital slot also simplifies the establishment of the communications link between a station and the satellite. Once the station's antenna is properly aligned, the antenna may only have to be repositioned to a significant degree when the station establishes contact with a satellite in a different slot. Prior to this era, a ground station's antenna had to physically track a satellite as it moved across the sky.

Based on these principles, three satellites placed in equidistant positions around the Earth can create a worldwide communications system in that almost every point on the Earth can be reached by satellite. This concept was the basis of Arthur Clarke's original vision of a globe-spanning communications network.

Uplinks and Downlinks. According to the FCC, an *uplink* is the "transmission power that carries a signal . . . from its Earth station source up to a satellite"; a *downlink* ". . . includes the satellite itself, the receiving Earth station, and the signal transmitted downward between the two."[1] To simplify our discussion, the uplink, for our purposes, refers to the transmission from the Earth station to the satellite, while the downlink is the transmission from the satellite to the Earth station.

This two-way information stream, the uplink and the downlink, is conducted with special equipment. The station that relays the signal must possess an antenna or dish that is usually parabolic in shape and a transmitter that produces a high-frequency microwave signal. Moreover, some ground stations both receive and transmit signals, while others operate in a receive-only mode.

The communications satellite, for its part, operates as a repeater in the sky. After the satellite receives a signal from the ground station, a signal is relayed back to Earth. This process is analogous to the function of an Earth-based or terrestrial repeater, but in the case of the satellite, the repeater is located more than 22,000 miles above the Earth's surface. The satellite

1. receives a signal from the Earth station;
2. the signal is amplified since it has lost a good part of its strength during its 22,300-mile journey;
3. the satellite changes the signal's frequency to avoid interference between the up- and downlinks;
4. the signal is relayed back to Earth, where it is received by one or more Earth stations.

To create this communications link, the satellite uses *transponders*, equipment that conducts the two-way relays. A communications satellite carries multiple transponders, and as illustrated by the Intelsat family, the

Figure 6–1
Artist's rendition of the Intelsat VI satellite. (Courtesy of Intelsat)

number of transponders per satellite class has increased over the years. This design evolution has led to the development of a new generation of satellites that can handle an enormous volume of information.

For example, the original Intelsat satellite, Early Bird, was equipped with 2 transponders that supported either a single television channel or 240 voice (telephone) circuits. The later Intelsat IV satellites improved upon this initial development. Each satellite, launched in the early to mid-1970s, carried 12 transponders that generally accommodated 4,000 voice circuits and 2 television channels.

The newer Intelsat VI spacecraft are equipped with 48 transponders. Intelsat VI can accommodate, on the average, "24,000 simultaneous two-way telephone circuits plus three television channels."[2] Using digital technology, a satellite could potentially handle 120,000 simultaneous two-way telephone circuits.

Intelsat VI is a hybrid satellite. It supports both the C and Ku-bands. As discussed later in this chapter, C and Ku-band satellites employ the C and Ku-band communications frequencies, respectively. Consequently, while some satellites employ only one band, other satellites support both, enhancing a satellite's communication capabilities.

Parts of a Satellite

Satellite Antennas. An important factor that influences the satellite communications network is the design of a satellite's antenna. A satellite's transmission is focused and falls on a specific region of the Earth, and this reception area, the *footprint*, can be shaped to conform to a country's particular geographical outline.

The Earth segment covered by a satellite transmission varies widely, depending upon the satellite and its projected applications. The footprints can range from *global* to *spot beams*, in descending order of coverage.[3]

A global beam provides the most coverage and is used for international relays. A spot beam falls on a narrowly defined geographical zone, which makes it particularly effective for covering major metropolitan areas. Since this signal is concentrated in a relatively small area, it is also stronger than one distributed to a broader region, as would be the case with a footprint that covers the country for cable-television relays. A smaller and less expensive dish could subsequently receive the spot-beam transmission within the confines of the reception area.

Satellite Spacing and Antennas. Geostationary communications satellites had traditionally been spaced four degrees or well over 1,000 miles apart in the orbital arc to create buffer zones between neighboring or adjacent satellites. The zones reduce the chance of cross-interference during transmissions. If the buffer zones were eliminated and satellites were placed too close together, an uplink that was not tightly focused and was intended for one satellite could unintentionally spill over and affect another satellite. Similarly, if an Earth station pointed its receiving dish toward a specific satellite, it could intercept a different satellite's signal.

In the 1980s, the FCC decided to reduce the distance between most communications satellites. Satellites would be situated closer together, on the order of two degrees, to open up additional orbital slots. This action was complemented by a set of technical specifications for Earth-based antennas to satisfy the demands imposed by the new spacing arrangement. These included a more accurate transmitting antenna that could relay a highly focused signal that would not spill into adjacent regions.

The FCC implemented a flexible timetable to cushion the impact of these decisions. A grace period was granted for the C-band satellites already in orbit and for the enforcement of the satellite-dish standards.

The plans were initially devised to address the orbital scarcity problem. A geostationary slot is the most advantageous orbital position for a communications satellite since the spacecraft can directly communicate with ground stations 24 hours a day. But the number of available slots is finite since the region of space that supports this type of orbit is confined to a narrow belt above the Earth.

A political consideration also plays a role

in the allocation of these highly valued positions. A nation cannot launch and place a satellite in any slot it chooses since the slots are assigned on an international basis through the auspices of the International Telecommunications Union.

The FCC's program was implemented to maximize the use of the orbital space allocated to the United States. It was also a reflection of the increased pressure to launch more satellites, which demanded the availability of additional orbital assignments.

Power System. The electrical power for the satellite is supplied through the conversion of sunlight into electricity by solar cells and ancillary equipment. Cylindrically shaped satellites, also known as spin-stabilized satellites, are covered with the cells. A three-axis stabilized satellite, in contrast, uses wings or extended solar panels for this operation.[4]

A communications satellite is also equipped with a battery backup system that is activated during solar eclipses, when the solar array is rendered useless. Some satellites may generate only enough battery power to operate at a reduced level, while others can function at full capacity until the eclipse ends.

Satellites also use another power source to remain operational during their average lifetime of 10 or more years. A satellite is equipped with external thrusters and a complementary supply of fuel. The thrusters, when activated by ground-station controllers, emit small jets of gas to help maintain the satellite's station, its position in its assigned slot. As indicated in the beginning of this chapter, a satellite in a geostationary orbit appears to be fixed in one position in the sky. In reality, the satellite moves slightly or drifts in its slot; the thrusters are employed to correct the drifting.

Once the fuel is expended after a period of years, the satellite is lost to ground operators.[5] The satellite's drifting can no longer be corrected, and the satellite may become inoperable even though its other systems still may be functional.

In view of this situation, the *Communications Satellite Corporation* (Comsat) devised a plan, the Comsat Maneuver, to extend the operational lifetime of select satellites. Under this system, a satellite would drift within a controlled area, and the satellite's fuel would not be expended at so rapid a rate to maintain its position. According to Comsat, a satellite's fuel consumption could drop from 37 pounds to only 3 pounds per year.[6]

The system's only drawback would be the modifications made to an Earth station that would communicate with the satellite. Since the antenna would track the satellite as it moved in the sky, the antenna must be equipped to carry out this procedure. Yet the upgrade would be compensated for by the satellite's extended useful lifetime. It would also reduce an organization's overall operating costs, since a satellite would not have to be replaced as soon as other typical spacecraft.

Finally, while communications satellites rely on solar energy to power their internal systems, spacecraft designed for deep-space and long-term missions have used *radioisotope thermoelectric generators* (RTGs), nuclear power. The RTGs have extended operational lifetimes and have provided power on missions where a spacecraft may be too far from the sun to tap its energy as a power source.[7]

Transmission Methods

A satellite's information capacity is limited by various factors, as are the other communications systems described in this book. In the case of a satellite, these include the number of transponders and the power supplied to the transmission system.

A typical 36-MHz transponder carried by a C-band satellite, for instance, can accommodate a television channel and a number of voice/data channels or subcarriers. As is the case with many terrestrial communications systems, satellite transmissions can be either analog or digital in form.

Various transmission schemes have also been implemented to enhance a satellite's communication capabilities, including the adoption of time and frequency multiplexing and *multiple-access systems*.

As the name implies, *multiple access* means that more than one ground station can access a satellite's transponder. Instead of dedicating the transponder to only one Earth station, it is used by multiple stations linked in a network.

This makes for an efficient transmission system. The information demands placed upon a ground station, may not necessitate the continuous use of the transponder's full channel capacity. Rather, the sharing of the transponder as needed may more accurately match the station's transmission requirements.

Two multiple-access systems have been frequency and time division multiple access (FDMA/TDMA). Due to technical considerations, a TDMA transmission has specific advantages over its counterpart. This includes the creation of a very efficient communications link that can use the transponder more effectively.

The primary objective of a TDMA and all other relay designs is to enhance a satellite's transmission capabilities in terms of its efficiency and user accessibility. Satellite traffic is also increasingly digital in nature, and digital-processing techniques have been adopted with the same goal in mind. These operations can also decrease the bandwidth requirement, as is the case with terrestrial networks, and lower the transmission costs.

An important implication of this development, digital compression, is making satellite communication available to a broader user group. By employing a compression system, an organization can use a portion of rather than a full transponder for a video relay.[8] While the receive sites must be equipped to handle this information, the transmission costs are reduced and a satellite becomes even more efficient, since it could carry additional channels.

This type of operation could handle a wide range of applications, including those outside of the traditional cable and broadcasting industries. In the education field, for example, a school may be able to start its own *distance learning network,* since satellite time would be less expensive.

A school could produce a series of educational programs, such as taped courses, that would be distributed by satellite and received by other schools in the network. If properly

Figure 6–2
The SpectrumSaver digital broadcast system. (Courtesy of Compression Labs, Inc.)

implemented, it could be an efficient and cost-effective way to share resources and to enhance a student's education.

It should also be noted, as indicated in previous and later chapters, that digital compression is not limited to the satellite industry. It is a powerful tool that has the potential to reshape whole segments of the communications industry.

Ku-Band

Until the early 1980s, transmissions in the commercial satellite communications field were primarily conducted in the 4/6 gigahertz (GHz) range known as the C-band. (The *4/6* notation indicates the downlink and the uplink, respectively.) C-band satellites are still the workhorses of the communications industry, and ground stations that receive their signals are generally equipped with satellite dishes that range between 10 and 15 feet in diameter.

Besides the C-band spacecraft, a newer generation of satellite that employs an even higher frequency range, the K-band, has become operational and is experiencing a tremendous growth. The first class of this type of spacecraft uses the Ku-band, with its 12/14 GHz frequency range.

One advantage of this higher frequency is the capability to upgrade the power of a satellite's downlink. In contrast with a C-band spacecraft, which uses 10 watts or less, a Ku-class satellite can operate at over 30 watts of power.

The C-band transmission is limited because it shares the same frequency range employed by terrestrial microwave systems, and the power of the downlink is restricted to avoid any possible interference. A Ku-band satellite, on the other hand, operates at a different frequency range, and thus the power of its downlink can be increased. This higher power translates into smaller receiving dishes and points out a generalization between the power of a satellite's transmission and a dish's size: as the power increases, the size of the dish can decrease.

The Ku-band also offers a dish owner more flexibility. A dish's smaller size and a Ku-band system's freedom from terrestrial operations simplifies the task of finding a suitable dish site. C-band systems are not afforded this same luxury: the possible joint interference and a dish's size may make it harder to find a location.[9]

Very Small Aperture Terminals. The development of Ku-band technology has spurred the growth of a relatively new type of satellite system that employs *Very Small Aperture Terminals* (VSATs). A VSAT is a compact satellite dish mated with the necessary electronic hardware to create an Earth station that can either receive or, in another configuration, receive and transmit signals.

The technology base is flexible in that a VSAT network can link a few, or if necessary, hundreds of sites. The actual network size is determined by the user and the company's particular needs. For an insurance company with branch offices located throughout the United States, it could create an extensive network tying all of its offices in this communications system.

As of the 1990s, VSAT technology has grown in popularity. The small dish can be mounted in a confined area, and a complete station is relatively inexpensive. A VSAT operation also supports a variety of network configurations, and a range of information, including computer data, can be relayed at different speeds. These characteristics, when combined with the capability to add new and even geographically remote sites to the system with relative ease, have made VSAT networks more flexible and less expensive than many land-based configurations.

Small organizations with limited budgets have likewise adopted this technology by pooling their resources. The companies install their own VSATs but share the expense of the master station, also known as the control facility or the hub. The hub is the heart of a VSAT network, and it is a sophisticated Earth station that routes and controls the flow of information through the system.

Ku-Band and Potential Problems. Despite the advantages brought about by the introduction of Ku-band satellites, as exemplified by VSAT networks, various sectors of the satellite communications industry in the United States, including the cable-television industry, are still heavily committed to C-band transmissions.

The Ku-band is more prone to interference created by rain than is the C-band. A severe rainstorm could attenuate the satellite's signal and deplete a portion of its energy.

At higher frequencies, as exemplified by the Ku-band, the signal can be weakened, that is, absorbed or scattered by raindrops.[10] This is analogous to the way the light from a car's headlights is dispersed and reduced in intensity by fog. Thus the transmission could be disrupted to a varying degree. This factor has been a consideration for organizations deciding whether or not to employ this frequency in geographical regions where heavy downpours are common.

SATELLITE SERVICES

The new developments in the satellite industry, such as more powerful transmission equipment and the opening of the K-band, have provided users with a flexible and effective communications system. This system has been further enhanced by multiple-access operations and by a wide channel capacity. The refinement of these various technologies helped bring about the revolution in satellite communication.

At present the U.S. communications satellite fleet consists of both government and commercial spacecraft. Government satellites are used for military and nonmilitary applications, including the creation of a worldwide communications net for the numerous military installations scattered around the globe. Other government-owned satellites, as demonstrated by the Advanced Communications Technology Satellite discussed in a later section of this chapter, are experimental in nature, and are designed to test new communications equipment and techniques. The satellite, in this particular case, is also a part of a comprehensive program that illustrates how the government and private sectors can work together in the satellite industry. Educational institutions, among others, have been encouraged to participate in this program by designing and eventually conducting experiments to evaluate the satellite's communication capabilities.

On the commercial front, Western Union's Westar I satellite, launched in 1974, was the country's first commercial satellite designed to serve domestic communication needs. Only two years earlier, the FCC, which oversees the private element of the satellite fleet, helped spur the industry's growth through its *open-skies policy*. The policy, in essence, encouraged the commercialization of outer space in regard to commercial satellite communication.

Finally, a company that plans to actually launch and operate a commercial communications satellite must first seek FCC approval. Other details must then be resolved prior to the launch, including contracting with an organization to place the satellite in orbit and obtaining insurance to protect the company against one of a number of possible losses. The losses could range from the destruction of the launch vehicle, and hence the satellite, to the satellite's failure to activate once it reaches its assigned station.

General Services

The satellite has emerged as a dominant force in the communications field since it is a reliable and relatively error-free communications tool. This is a reflection, in part, of the integration of digital components and trans-

missions. A satellite also supports a high-speed data link, and this application highlights a growth area in business communication. More companies are taking advantage of a satellite's wide channel capacity to relay computer data and other forms of information, including digitized audio and video signals. This is in keeping with the current trend toward developing integrated streams that carry a mixed bag of information via a single transmission conduit.

A satellite becomes an even more potent tool when examined in light of its multipoint distribution capability. A number of sites can be linked by a videoconference, an electronic meeting, and rural areas not supported by terrestrial lines can receive satellite transmissions with the appropriate receiving dishes.

These different satellite services have been and will continue to be offered on both a national and international basis, and numerous companies either provide turnkey satellite communications systems or lease satellite channel space to organizations that have their own satellite dishes. Accordingly, a company that wishes to use a satellite as a communications vehicle can lease a transponder for a specified fee and for a set time period. Another option calls for leasing time for an occasional relay through an intermediary, an organization that subleases transponder time.

Intelsat, Inmarsat, and Comsat. The international market, at this time, is dominated by the *International Telecommunications Satellite Organization* (Intelsat). Founded in 1964 and composed of more than 100 nations, including the United States, Intelsat is the world's largest satellite consortium. Intelsat owns and operates a fleet of satellites that supports a list of services ranging from the international distribution of television programming to the relaying of telephone signals.

Through another organization, the *International Maritime Satellite Organization* (Inmarsat), ships at sea, oil-drilling rigs, and even remote sites on land have been supported by a satellite communications network. A ship, for example, could establish a link with a satellite to communicate with a land base. In this particular operation, a special stabilized antenna, which can maintain its bearing toward a satellite even in heavy seas, has been used.

Comsat, for its part, has served as the U.S. representative to both organizations. This has made Comsat a powerful and far-reaching satellite organization that provides an array of services.

But there are storm clouds on the horizon. Independent organizations, such as PanAmSat, have stepped forward to directly compete with Intelsat in the international satellite market. While other preexisting satellite networks have served various regions of the Earth, they are generally not viewed as competitors to Intelsat's operation. The new private networks would, however, be direct competitors for satellite communication traffic.

The actual creation of private satellite networks has been made possible, in part, through the rapid integration of satellite technology and equipment in the world market. When Intelsat was founded, the communications satellite industry was in its infancy, and the consortium was established to foster international satellite communication for both developed and developing nations.

Today, the satellite is an established element of the world's communications system. The tools of satellite technology have reached a level of development and cost-effectiveness where it is now possible to establish private satellite networks, as witnessed by the proliferation of VSAT systems in the United States. In an analogous fashion, it is possible to develop private international networks that would complement the Intelsat system.

Teleports. Satellite networks, be they national or international in scope, can work with terrestrial communications systems. A company's data may be transmitted over a high-speed land line prior to an uplink and after the downlink. Consequently, satellite and terrestrial systems can be interdependent. The satellite transmission can serve as the long-distance connection, while terrestrial lines provide the intracity hookup.

This integration of systems has been exemplified by New York City's *teleport*. The Port Authority, Western Union, and Merrill Lynch joined forces to develop and operate a sophisticated communications center on Staten Island. A series of ground stations has tied the teleport to national and international satellites while high-speed lines have provided the local connection for metropolitan New York and parts of New Jersey.

According to a teleport consulting firm, the ELRA Group, 150 teleports will be operational in North America by the turn of the century.[11] The sites, in tandem with a variety of terrestrial networks, will serve the growing needs of the business community through the transmission of voice, data, and video-based information.

Teleports will also help spur the growth of the satellite industry. A number of organizations could share various satellite facilities to reduce each participant's financial burden. This could make satellite communication more economically attractive for organizations that have not yet entered the field.

Similarly, a company may not be able to construct an Earth station due to its geographical location, such as the tenth floor of a building in midtown Manhattan. The establishment of a nearby teleport would provide the company and other organizations in the same situation with the capability to establish a satellite link.

Satellites and the Broadcast and Cable Industries

Satellite-distributed television programming is the backbone of the cable-television industry in the United States. A company such as HBO uplinks its programming to a satellite, where it is subsequently downlinked and received by cable companies throughout the country. The programming is then locally distributed to individual subscribers via cable. It was, interestingly enough, the placement of HBO on satellite that helped spur growth of cable systems in urban areas.

A number of independent television stations, including WPIX in New York City and WTBS in Atlanta, have similarly joined the satellite revolution. These stations, also called Superstations, use satellites to distribute their programs much the same way as HBO.

The three major television networks have also turned toward satellite communication. NBC, for example, the first Ku-band network, established a national satellite system that linked affiliate television stations, while ABC used a satellite to distribute nationally its radio programming, among other operations. These developments follow up on the success enjoyed by the Public Broadcasting System, one of the pioneering organizations that helped establish the satellite as a potent communication force in the broadcast and cable industries.

Satellite Newsgathering. The growth in satellite communication has revolutionized another facet of the television industry: television news operations. Besides being able to relay and receive national and international news at a moment's notice, television stations can now participate in *satellite newsgathering* (SNG), a newer form of production.

The availability of satellite channels and the lower price for portable satellite dishes and other components have made it possible for an individual television station to create its own remote satellite uplink. In a typical operation, the station buys either a van or a truck and a portable satellite dish. This rig can then be taken on the road, and when the reporters reach the site of a story, an uplink to a satellite is established. The transmission, the story, is subsequently picked up by the home television station through a downlink.

This form of reporting has emerged as a powerful communications tool. A station can conduct an SNG relay from a site that may have been too far removed from the station in the past to initiate this type of operation. For example, a station from Seattle, Washington, could send its satellite rig to Washington, D.C., to establish a link between Seattle's congressional representatives and their constituents back home.

Satellite newsgathering systems are just beginning, relatively speaking, to have an impact on the broadcast and cable industries. Television stations are now providing their viewers with information that may have been unavailable in the past, while other stations are forming mininetworks to ex-

Figure 6–3
The S23 Ku Uplink vehicle. A reporter can relay stories back to the home station through such a mobile link. This development helps break the physical restrictions that prevented reporters from relaying live stories that were geographically removed from the station. (Courtesy of Harris Allied Systems)

change stories that may be applicable to different groups.

On the international front, special *flyaway systems* have extended satellite communication to regions where standard satellite links may not be available or readily accessible. A flyaway is a portable system that can be flown on a commercial airliner. The system is stored in trunks and reassembled upon arrival. The Cable News Network (CNN) has pioneered the use of such systems to cover world events, including the 1989 student uprising in Beijing, China, and the Persian Gulf conflict, Operation Desert Storm, in 1991.

Satellite technology also made Operation Desert Storm the first "real-time" war.[12] People around the world witnessed events, such as Scud-missile attacks, as they were actually occurring. Satellite links also provided reporters with the ability to relay voice, video, and computer data, news stories, back to their home offices in a timely fashion.

This immediacy triggered negative and positive responses in the United States. For example, Peter Arnett, a CNN reporter, continued to file stories from Baghdad, Iraq,

Figure 6–4
A flyaway configuration. The capability to pack a satellite communications system and fly it to various points around the world has helped revolutionize international news coverage. (Courtesy of and © CNN, Inc. All Rights Reserved.)

during the war. He was criticized for this action by groups such as the Victory Committee, a coalition that included the Accuracy in Media organization.[13] The criticism centered on his reporting while under Iraqi censorship and the fact that stories were used

for pro-Iraqi propaganda purposes. Similarly, though, severe restrictions and constraints were placed upon the press by the U.S. government and its allies.

While satellites and other communications tools made it possible to monitor the Gulf War to an unprecedented degree, the use of new technologies does not guarantee good reporting. As stated in a report that explored the media's role in this conflict, "technology cannot be an end in and of itself in making sense of war-related events. That, as always, remains the job of the journalists themselves, with the new technology facilitating but not replacing the task."[14]

Remote Sensing and the News. Pioneered in part by Mark Brender and ABC News, the electronic and print media have adopted another satellite-based system, *remote-sensing satellites,* to enhance their news coverage. A remote-sensing satellite is a sophisticated spacecraft equipped with an array of scientific instruments, including high-resolution cameras. Instead of being locked in a geostationary orbit where the same restricted geographical area is surveyed, the satellite covers the whole Earth in successive orbits. The pattern is then repeated.

Remote-sensing satellites were originally designed to examine and explore the Earth. They have been used to document the Earth's physical characteristics and locate oil deposits and other natural resources. The United States and France have led the world in this field through their *Landsat* and *SPOT* satellites, respectively, and news organizations have subsequently used this tool to enhance their coverage.

For example, this type of satellite proved to be an invaluable news resource in the aftermath of the Chernobyl nuclear reactor accident. Satellite pictures of the site were obtained and released by the media. The pictures highlighted the true extent of the facility's damage and helped prevent a potential cover-up of the situation.

In another setting, ABC News used remote-sensing images in a special program televised in July 1987. The pictures highlighted various facets of the Iran-Iraq war. The same region was also under scrutiny

Figure 6–5
Artist's illustration of Landsat 6. (Courtesy of the Earth Observation Satellite Company, Lanham, Md.)

LANDSAT 6 SPACECRAFT

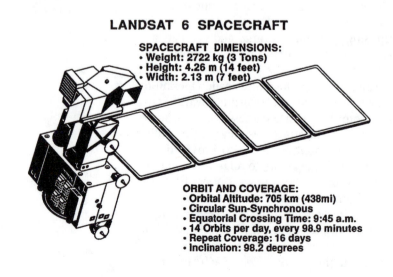

SPACECRAFT DIMENSIONS:
• Weight: 2722 kg (3 Tons)
• Height: 4.26 m (14 feet)
• Width: 2.13 m (7 feet)

ORBIT AND COVERAGE:
• Orbital Altitude: 705 km (438mi)
• Circular Sun-Synchronous
• Equatorial Crossing Time: 9:45 a.m.
• 14 Orbits per day, every 98.9 minutes
• Repeat Coverage: 16 days
• Inclination: 98.2 degrees

during Operation Desert Storm, when pictures depicted the conflict area.

Despite the benefits offered by this type of service, all governments, including the U.S. government, have not been happy with this newfound capability. Since the media could order a photograph of a region of the Earth covered by the spacecraft, military maneuvers could be documented and other situations a government may want to keep hidden from the general public could be revealed. In the case of the United States, the government could have imposed restrictions on private, domestic remote-sensing licenses, via ambiguous licensing procedures. If implemented, the media's access to specific images may have been affected and limited.[15]

This stance, and others, triggered a reaction from various media groups, including the Radio-Television News Directors Association. The association argued that the restrictions violated First Amendment rights and freedoms, and that another government concern, the potential violation of national security, was unfounded. The press had generally been responsible in the use of sensitive information in the past and would continue to do so with the satellite pictures.

Besides this initial government intervention, media organizations were and still are faced with other major problems that may hinder the use of this investigative tool.

First, a satellite may not be in the correct orbital position to immediately deliver a picture of a specific region of the Earth. Consequently, a news department may have to order a picture from a vendor, such as the Earth Observation Satellite Company (EOSAT) or the French SPOT Image Corporation, the operators of the Landsat and SPOT satellites, respectively, several days before a telecast. This delay could hamper the coverage of fast-breaking news events.

Second, a picture relayed by a satellite is not in the form of a standard video or still image. The picture must be processed before it is appropriate for a television audience, and an expert may have to be called in to provide a correct analysis.

The former problem has been somewhat alleviated, though, by the introduction of software that simplifies the manipulation of remote-sensing images. ABC News, for one, employed such a system during Operation Desert Storm to generate graphics for its programs.

Another solution to the overall problem may be offered in the form of a *Mediasat,* a remote-sensing satellite(s) specifically designed for the media. The satellite could carry high-resolution instruments and could possibly photograph various regions of the Earth on shorter notice. It may also be possible to use a smaller, less expensive satellite, as described in a later section of this chapter, to set up the system.

Yet the cost to build and operate a spacecraft, as well as the hiring of a support staff, may still be prohibitive. There are also some unresolved issues, at least for U.S. companies that may want to engage in this venture, even though the original restrictions have been generally relaxed.[16]

Satellites and the Scrambling Issue. Both independent television stations and specialized entertainment services, such as WTBS and HBO, are distributed by satellite to cable companies across the United States. Once the signals are received, they are routed to individual subscribers through local cable operations.

This multipoint distribution system is very efficient, and a single satellite can simultaneously relay a program to hundreds of cable companies. But the same capability creates a problem for the industry. Since a satellite transmission for this type of application has a broad footprint, anyone who owns a satellite dish can also receive the satellite downlink and the programming. This would enable an individual to bypass the local cable company and to gain access to HBO and other services for free. Also individuals could retrieve other satellite-distributed programming, such as paid videoconferences.

This situation developed into a serious problem by the mid-1980s. At this time, the *Television Receive-Only* (TVRO) industry, partly composed of companies that manufacture and supply Earth stations, experienced a rapid growth in the consumer end of the business. This was a result of the FCC's 1979 deregulation of receive-only stations that made a previously required licensing arrangement optional.

Consumer-based TVRO systems, consisting of small backyard receive-only dishes and the complementary electronic components, were purchased by more than a million Americans. These configurations, which can be called *home satellite dishes* (HSDs), sprouted up virtually overnight in metropolitan and rural areas, and people gained access to and watched pay-television services and other satellite-distributed programming for free.

As the number of HSDs continued to rise, both cable systems and program providers realized they were losing money. At this juncture, the VideoCipher encryption or scrambling scheme was developed and widely adopted by organizations such as HBO.[17]

A television signal was rendered unintelligible by the system unless an individual possessed a special descrambling device that restored a program to its original state. In the case of cable companies, each site would be equipped with a descrambling unit so legitimate subscribers would continue to receive uninterrupted programming.

Full-time scrambling was inaugurated in 1986 by HBO, and then other satellite-distributed services joined the bandwagon. This development created a furor in the TVRO industry. Sales for HSD systems declined. Dish owners had two alternatives: either buy or lease a descrambling device for several hundred dollars and pay a monthly subscription fee for receiving one or more of the now-scrambled television channels; or avoid the monthly fee and use an illegally altered decoder. Illegal decoders became available when the supposedly unbreakable scrambling system was cracked.

One individual became so angry at this whole situation that he illegally interrupted HBO's programming on April 27, 1986, and relayed his own antiscrambling and antisubscription-fee message:

Good evening HBO
From Captain Midnight
$12.95/month
No way!
Showtime/Movie Channel beware

This satellite pirate, Captain Midnight, was eventually identified as a part-time employee at a teleport and was subsequently convicted.[18] Captain Midnight used the teleport's facilities to override HBO's signal, and the satellite distributed his signal, the message, instead.

Each group of players—HSD owners and the TVRO industry on one side and the cable and television industries on the other—has presented its own arguments in support of its position. Scrambling opponents indicated that many backyard dishes were purchased by individuals who were not served by cable companies and over-the-air broadcast stations. Their only recourse was to buy an HSD system, and this expensive solution became even more expensive with the introduction of the scrambling device and one or more subscription fees.

The television and cable industries, for their part, asserted their property rights for the programming: illegally received signals were pirated signals. They also stated that some individuals purchased a dish to avoid signing up with a local cable company.

The solution to this problem lay in a compromise. Companies lowered their prices for subscriptions, and HSD owners were generally treated more equitably, on a par with standard cable subscribers in terms of monthly subscription fees. These initiatives, among others, helped to somewhat defuse the situation, but the problem with illegal descramblers continued.

As with computer software, the philosophical basis behind the idea of ownership and property rights—in this situation, that of satellite-distributed programming—has to be accepted. Otherwise, legal and technical measures must be adopted. But in the technical arena, even enhanced protection schemes can potentially be broken.

The General Instrument Corporation, the VideoCipher's parent company, has campaigned against illegal descramblers for a number of years. But estimates indicate that at one time, a quarter of the crop of descramblers were not authorized to receive programming.[19]

In an attempt to solve this problem, an enhanced version of the decoder was adopted in the early 1990s. It was the next logical step in what could be a series of steps, to ensure a satellite relay's integrity.[20]

Finally, the controversy between these two groups has clouded another important issue, the vulnerability of the commercial satellite fleet. If Captain Midnight could disrupt HBO's transmission, other individuals with access to the proper facilities could follow suit. Even though this group may be limited in size at this time, the situation could change as new facilities, both permanent and portable, are brought on-line.

Beyond television programming, information vital to the operation of the country and the world, such as financial data, is relayed on a daily basis. An ongoing disruption of these services would be disastrous. Ultimately, the same technology that advanced and enhanced our communications system may harm the entire country and the international community unless the necessary precautionary steps are taken.

This problem has so many ramifications that the FCC indicated that various security systems had to be implemented to protect the integrity of satellite transmissions. In 1991 the FCC took a step in this direction by adopting an automatic transmitter identification system (ATIS). Designed for video broadcasts, "ATIS repeats the name and phone number of the broadcaster in Morse code on a subcarrier just beyond the audio spectrum, making signal identification easy."[21]

The system was implemented so that accidental as well as intentional interference sources could be quickly identified. But a similar system has yet to be adopted, at least at the time of this writing, for nonvideo relays.

DIRECT-BROADCAST SATELLITES

Domestically, a *direct-broadcast satellite* (DBS) is, as of this writing, a proposed class of spacecraft with very powerful transmission systems, on the order of 150 to 200 watts of power. A satellite would operate in the K-band and, as originally conceived, would bypass television stations and cable companies to relay programming to consumers equipped with special receiving dishes. A DBS operation could be viewed, in this context, as an enormous bypass operation, much in the same vein as a VSAT system.

An important element of a DBS operation would be the use of small satellite dishes. Thanks to a direct-broadcast satellite's downlink, which would exceed the power of a conventional satellite's relay, a dish less than two feet in diameter could easily receive the signal. This is in contrast with the much larger dish, on the order of seven or more feet, employed in a typical C-band HSD configuration. The dish size would be a vital concern for DBS companies, since a small dish would be unobtrusive, compact, fairly inexpensive, and easy to set up. The dish would also be permanently aligned and fixed toward a specific satellite when installed, unlike an HSD dish, which can be moved and pointed at different satellites.

A DBS company could provide subscribers with programming ranging from television shows and movies to specialized information services. The same operation could also support an enhanced television relay with high-quality pictures and sound. But, as examined in Chapter 2, it is difficult to introduce sweeping changes in the traditional broadcasting field. Any significant and rapid technological changes would result in the potential obsolescence of the equipment owned by consumers and producing organizations.

A DBS operation could help solve this dilemma. Only those individuals who elect to receive a new service would purchase the equipment necessary for its reception. Consequently, an incompatible television transmission system, which generates higher-resolution pictures, could be implemented through a DBS service without affecting the entire country. Other nations have adopted this stance, while the situation in the United States, as described in Chapter 9, is different.

DBS Systems—An Evolution

The concept of what constitutes a DBS service, at least in the United States, has evolved over the years. The vision of a system of very high power satellites has shifted, in some quarters, to an alternate view. A less powerful spacecraft, a medium-power Ku-band satellite, has been promoted as a replacement for the traditional DBS model, which failed to emerge as a viable commercial enterprise for various reasons.

First, the development of a nationally supported DBS system demanded an enormous capital investment. Beyond the millions of dollars to build, launch, and maintain the satellites, it was necessary to create a comprehensive terrestrial support network. This included items such as local sales and repair offices, an advertising campaign, and program-licensing fees.

Second, the rapid expansion of the TVRO/HSD and VCR industries exacerbated this situation. A large segment of the consumer market targeted by the DBS industry was already served by HSD systems and VCRs. More than 40 million households were already equipped with VCRs by the mid-1980s, and this phenomenal growth rate made it possible for Americans to view a diversity of programming, especially movies.[22] This was an unfortunate development for DBS companies, since movies were slated to be one of their staple features.

Third, the DBS industry could not attract and sustain a sufficient level of financial support. It was also dealt a severe blow in the 1980s when the Satellite Television Corporation, a subsidiary of Comsat, suspended its plans to inaugurate a DBS system. Other companies, including CBS, had previously bowed out of the field.

Consequently, the high-power DBS system did not materialize in the United States during the 1980s. This development gradually led to a revision of the original idea for direct-broadcast services. Rather than creating a fleet of expensive and untested high-power satellites, a more cost-effective and proven satellite, a medium-power Ku-band spacecraft, could be used to initiate a "low-power" DBS service.

In the latter half of 1983, while the DBS industry was undergoing a series of upheavals, a low-power operation was created by United Satellite Communications, Inc. (USCI). It was a commercial venture that offered subscribers five channels of entertainment programming. Future options tentatively included specialized information services, such as videotex and data transmissions, as well as bilingual programming.

Despite the advantages of using a less expensive spacecraft and beating the proposed high-power operations in developing a working communications system, financial pressures forced USCI to close its operation in 1985. But the company did demonstrate the feasibility of starting a DBS service without one of the high-power DBS satellites.

But even a low-power DBS operation would have to contend with some of the problems that affected its high-power counterpart, including the saturation of VCRs in the consumer market. A low-power satellite would also be at a technical disadvantage with respect to its weaker signal strength and the superior orbital spacing arrangement designated for the higher-power spacecraft. High-power DBS satellites would be spaced farther apart than Ku-band satellites, and this buffer would benefit individuals on Earth using small receiving dishes since the potential for interference between adjacent spacecraft would be reduced.

More Recent Plans. As of the early 1990s, other DBS ventures are still on the drawing board or have gone the way of USCI. One ambitious proposal, initiated by NBC, Hughes Communications Inc., Cablevision Systems Corporation, and the News Corporation Limited, was a high-power DBS system. It was unique in the diversity of its partners and the adoption of a digital-delivery system. The plan collapsed, however, in the early 1990s.[23]

Despite this additional setback, other companies are still pursuing the DBS golden goose. These include Hubbard Broadcasting, whose president, Stanley Hubbard, has been one of the most vocal DBS advocates in the United States.

Hughes Communications, Inc., continued its interest in the DBS field and helped plan a new service, DirecTV, that could deliver more than 100 digital channels to subscribers via high-power DBS satellites. The service would support pay-per-view options, such as movies, in addition to other, more standard programming fare.[24]

The 100-plus channel capability, made possible through digital-compression techniques, is also an important feature. Unlike earlier proposals designed to support a limited number of channels, DirecTV could directly compete with cable systems' programming options.

The attraction of a DBS system is a powerful one, which fuels the continued interest in the field. As indicated, consumers could receive a wide array of programs that may not be otherwise available. The prerequisite technology base has also matured since the 1980s, making DBS operations more feasible.

A number of scenarios could develop in this market. DBS operations may, for example, become fully integrated in the United States, especially if enhanced television services are not made available in a timely fashion over traditional terrestrial channels. DBS operations could also complement the existing communications infrastructure. In some cases, individuals could supplement their current programming options.

But the same problems that plagued the industry in the 1980s could have a similar impact in the 1990s. There is an additional wrinkle in this scenario: fiber-optic systems. As described in Chapter 5, an FO communications system would enable subscribers to receive a diversity of programming. Cable companies are laying new lines, and the telephone industry has an interest in this field. These developments could make FO and DBS systems compete for programming and subscribers.

Or, as is the case with other elements of the communications industry, DBS satellites and FO systems may actually reach a balance where the technologies and applications coexist. But the potential form of this communications structure is still unclear.

Finally, the DBS picture is remarkably different in other parts of the world. A number of countries are well versed in DBS operations and continue to draft plans for sophisticated DBS plants. Japan and various European nations are the major contenders in this field. As other countries' DBS projects continue to mature, the United States may be forced to follow suit with its own high-power DBS configurations to keep abreast of emerging technologies in the international market.

FUTURE SATELLITE TECHNOLOGY

Besides the proposed direct-broadcast satellite, the plans for the next generation of conventional communications satellites have already been drafted. The spacecraft will improve upon current designs and will carry sophisticated on-board switching and processing equipment. These intelligent satellites will direct the flow of communications signals at high bit rates, which will help simplify the ground network's design and will make it easier to establish a station-to-station link. Ultimately, such satellites will reduce the cost to create, run, and maintain the overall satellite and terrestrial communications system.

This new generation of spacecraft is exemplified, as previously indicated, by NASA's *Advanced Communications Technology Satellite* (ACTS). ACTS will use the Ka-band, another subset of the K-band, for transmission purposes. The Ka-band has a higher frequency than that employed by Ku-class spacecraft and, like the Ku-band, is adversely affected by rain. But the ACTS system will be served by an enhanced monitoring and correction scheme to compensate for this problem.

Besides these attributes, ACTS will be equipped with a multibeam antenna. As part of a sophisticated transmission configuration, the antenna system was initially designed to support fixed and hopping-spot beams.[25] A tightly focused and fixed beam would be stationary and pointed at the same geographical region. A hopping-spot beam, that is, a scanning and a movable beam, would literally hop to different regions of the country. ACTS would receive or relay data through abbreviated ground links at these specific sites, and the beam would then move to other locations.

All in all, ACTS is a prototype for the future communications satellite fleet. It will sim-

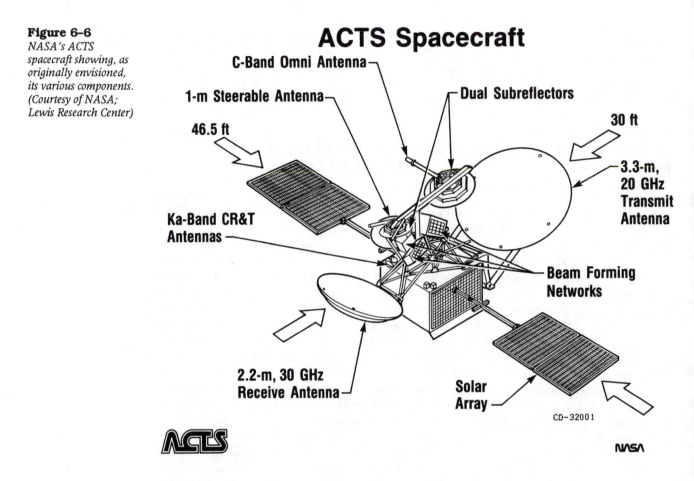

Figure 6–6
NASA's ACTS spacecraft showing, as originally envisioned, its various components. (Courtesy of NASA; Lewis Research Center)

plify the ground network and will enhance a satellite's relay capabilities through the development of an advanced transmission system (e.g., the hopping beam can be configured to match a region's communications demands).

ACTS was originally designed to accommodate an experimental optical communications system. Even though this particular system, which was devised by the military to produce a secure communications relay, was scrapped from the mission, it may be incorporated in future satellites. A configuration could be used for a satellite-to-satellite communications link and possibly to create a high-speed satellite-to-ground relay.

A final proposal that will carry the development of satellite technology one step further is the plan to create and launch a series of space platforms instead of standard communications satellites. As envisioned, a space platform would be a large structure placed in orbit that would route an immense quantity of information through multiple transmission systems and antennas. The satellite would occupy a single geostationary slot and could potentially replace several or more contemporary spacecraft. The development of space platforms could also open up additional slots for other applications, since fewer spacecraft would have to be launched to support the satellite communications net.

LAUNCH VEHICLES

The growth of the satellite system will be fueled by the creation of new launch vehicles and the entry of new organizations into the field. In the past, companies and most nations that wanted to launch a satellite had to sign on with NASA. But the large number of satellites placed a heavy demand on launch-vehicle and facility availability.

It appeared this situation was going to be altered in the 1980s. The entry of a private consortium in the launch industry, *Arianespace*, in addition to the development of NASA's *space shuttle*, promised to facilitate satellite-launch operations.

The Space Shuttle

The space shuttle is the world's first refurbishable manned spacecraft. After a mission, the shuttle returns to Earth and is refurbished in preparation for its next flight.

The space shuttle can carry a wide range of cargo within its hold, the cargo bay, including a self-contained laboratory, scientific experiments, and satellites. In its latter role as a satellite launch vehicle, the shuttle initially carries a satellite to a low Earth orbit. At this point, the satellite is released from the hold by a special mechanism.

If the satellite's final destination is a geostationary orbit, a rocket booster attached to the satellite is activated. This booster propels the satellite to a specified altitude, and it eventually reaches a preassigned orbital slot after a series of maneuvers.

Despite the shuttle's promise as the first of a new generation of spacecraft that would make this type of space-related activity an everyday event, the shuttle was plagued by mechanical and structural problems. This was partly a reflection of the spacecraft's heritage. The shuttle was created to accommodate a broad range of activities, and since it was a refurbishable as well as a manned vehicle, its design was complex in comparison with a standard unmanned rocket, an *expendable launch vehicle* (ELV). This complex design, when combined with other factors, inevitably led to a series of problems. In one instance, special tiles that protected the shuttle during its reentry into the Earth's atmosphere had to be replaced more frequently than had been originally anticipated. This delayed the refurbishing process and the spacecraft's next possible launch date.

The shuttle program was also hampered by a burdensome organizational hierarchy. The spacecraft was developed as a joint venture between NASA and the military, and military payloads had priority over civilian payloads. This had an impact on the commercial satellite industry in respect to securing specific launch dates.

Setbacks to the Program and the *Challenger* Explosion

In February 1984 the space shuttle's credibility as a launch vehicle received another blow when two satellites failed to achieve their assigned orbital positions. After the Westar VI satellite was released from the shuttle's hold, its rocket booster was supposed to propel it into an elliptical transfer orbit. Another small rocket should then have placed Westar VI in its final geostationary slot. But the booster malfunctioned, and the satellite was placed in a useless orbit. Indonesia's Palapa-B2, the second satellite launched during the mission, suffered a similar fate.[26]

A little less than two years later, the world was shocked by the explosion of the space shuttle *Challenger*. The entire crew was lost in the most devastating tragedy in NASA's history.

In the wake of this disaster and a report released by a commission convened to investigate the explosion, then-President Ronald Reagan announced that NASA would generally withdraw from the commercial satellite launch industry other than fulfilling an undetermined number of preexisting contracts. The frequency of future flights would also be scaled down, and the shuttle's primary role would be to support scientific and military missions.

This directive reflected the president's attitude toward the government's role in private enterprises and the realization that an overly ambitious launch schedule contributed to *Challenger*'s destruction. As stated by the commission, "The nation's reliance on the Shuttle as its principal space launch capability created a relentless pressure on NASA to increase the flight rate."[27]

This pressure played a role in the decision to launch *Challenger* under adverse weather conditions. Other contributing factors, which directly led to *Challenger*'s destruction, were design flaws in the shuttle's rocket boosters, as well as possible flaws in the spacecraft's overall design and the booster-refurbishing process.[28]

A number of companies have subsequently stepped forward to fill the void in the commercial-launch industry. Even though there was some activity in this area prior to President Reagan's announcement, the list of interested parties has grown at an accelerated rate.[29]

Existing ELVs as well as newer designs have been adopted, including the development of such rockets as the *Pegasus,* a launch vehicle developed by the Orbital Sciences Corporation. Instead of the typical launch from the ground, Pegasus is carried by a jet, released, and then proceeds on its own power.

Developed as a relatively inexpensive launch system, Pegasus could carry small payloads into low Earth orbits. Its capabilities have also been matched by a new generation of *lightsats.*

Lightsats are small, inexpensive satellites that could be used for remote sensing, the potential creation of cost-effective personal communications networks, and for other applications.[30] The importance of this development, the bottom-line figure for countries, institutions, and possibly even news organizations, is that space has become more accessible.

A lightsat would be less expensive than a conventional satellite and could be designed and assembled in a shorter time period. It would then be launched on a rocket specifically tailored for small payloads rather than using a more expensive, standard ELV.[31] The trade-off is that the lightsat may not be as sophisticated or may not accommodate as many tasks as its larger and more costly counterpart.

Consequently, during the next few years, the American satellite launch fleet will include the space shuttle and various ELVs. The commercial sector will use ELVs to carry payloads into orbit while the government will rely on the shuttle and ELVs.

NASA has helped support this industry by leasing its facilities to private companies and through other programs. This support is vital in view of the stiff competition American companies have faced and will face in their search for a share of the international satellite launch market.

This mixed fleet of space shuttles and ELVs has also provided the United States with a more balanced launch capability. When the shuttle program was in full gear, ELVs were delegated to a secondary role. But they have reemerged from the background.

The space shuttle, for its part, will continue to fulfill two of the roles for which it is best suited, those of a research and a special utility vehicle. An example of the latter was a dramatic, televised satellite salvage operation that took place in a low Earth orbit.

In 1992 the shuttle *Endeavour* rendezvoused with an Intelsat VI satellite that failed to achieve its final orbital position. After some difficulties, three astronauts eventually helped retrieve the satellite and brought it into the shuttle's bay. A motor was subsequently attached and the satellite was released for a boost to its final geostationary position.

Besides contributing to the knowledge base for recovery operations of this nature, the mission revealed some of the potential shortcomings of simulations when applied in real-life situations. A simulation, which attempts to duplicate the conditions of an actual event, was inaccurate in this case. An alternate plan had to be devised and sub-

Figure 6–7
Pegasus launch vehicle. (Courtesy of Orbital Sciences Corp.)

sequently carried out to complete the mission.

This type of practical experience, of dealing with novel situations, is important for the future of extravehicular space-based activities. It also highlighted, especially at this stage in the development of our technology base, the importance of the human presence in space. Humans, unlike current robotic devices, have the flexibility to adapt to unique circumstances, and in this particular case, to complete the mission.[32]

Arianespace

Arianespace is a private commercial enterprise and an offshoot of another European organization, the *European Space Agency* (ESA). Arianespace was created in response to NASA's earlier domination of the satellite launch industry.[33]

Arianespace has aggressively promoted and marketed a series of ELVs and a sophisticated launch-support operation. At this time, the organization's Ariane rocket series can accommodate a range of payloads, and its launch site in Kourou, French Guyana, is particularly well situated to place satellites in geostationary and other orbital positions.

Arianespace has also devised and maintained a competitive price structure and a reliable launch schedule. These factors contributed to its growing share of the international launch market when NASA was still a participant in the field.

Despite its successes, Arianespace has suffered a number of its own setbacks. Satellites have been destroyed as a result of rocket failures, and the organization must face a host of new competitors, including China, the United States, and Japan.

The Current and Future State of the Satellite Launch Industry

The state of satellite launch technology is a vital concern to the communications industry. But we tend to overlook the role and contributions of this field and solely concentrate on the satellites and their applications. This may be a short-sighted view, since the launch and satellite industries are intrinsically linked.

The commercial launch industry, as indicated, experienced a series of upheavals during the mid- to late 1980s. While it is true that NASA had left the field, a number of companies and nations filled the void. In fact, the increased competition triggered by NASA's decision may make it easier for an organization to launch a satellite. There are more companies and a greater supply of ELVs on the market.

However, on a bleaker note for the United States, NASA's departure from this field could have a devastating impact on the country's launch capability if independent companies cannot compete with other nations and commercial enterprises, such as Arianespace. This situation is only exacerbated by the recent and growing erosion of the United States' "preeminent position" in the satellite-manufacturing field. During the late 1980s to early 1990s, 36 communications satellites were manufactured by the United States and 23 were built by Europe and Japan.[34] Prior to this time, the United States dominated the industry.

Finally, one of the original goals of the shuttle program, to make space affordable and accessible, may be satisfied through the *National AeroSpace Plane* (NASP) Program and

Figure 6–8
The new generation of ELVs may help launch enhanced communications systems, such as a multisatellite network that would provide personal communications services. This particular service, Iridium, is described in Chapter 11. (Courtesy of Motorola Inc.)

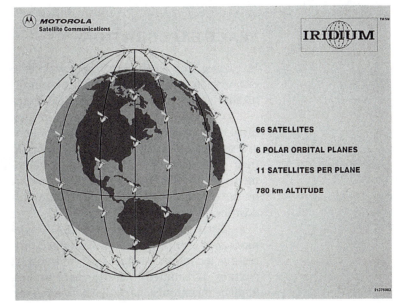

its experimental X-30 flight vehicle. The X-30 could pave the way for a new series of aerospace planes that could take off and land on conventional runways, attain a low Earth orbit, and be *reusable* instead of *refurbishable*.[35]

Reusable implies a quick turnaround time and the ability to prepare a vehicle for its next flight without major refurbishing, unlike the space shuttle. This capability would save time and money.

The projected applications for future aerospace planes based on the X-30 are many. Satellites in a low Earth orbit could be retrieved and, if necessary, repaired. A vehicle could also serve as a ferry to bring supplies and personnel to and from *Freedom*, NASA's proposed space station. Another spinoff is more down to Earth, literally. The NASP program may promote the development of a new generation of fast commercial airliners that could, for instance, carry passengers between Los Angeles and Sydney, Australia, in 2.5 hours instead of the usual 13.5 hours.[36]

Other countries are in the process of designing their own vehicles. If the technology and the resulting designs reach their full potential, aerospace planes from different nations will both complement the new, mixed launch fleet and will further promote commercial enterprises. Just as important, they may help make space accessible to all of us.[37]

SPACE EXPLORATION

In closing this chapter, it's appropriate to examine an area related to satellite communication: space exploration. Both fields coincide to a certain extent, and the space probes, fitted with imaging and nonimaging systems, are sophisticated communications tools in their own right. More important, developments in this field have had a direct impact on the communications and information industries.

As discussed in Chapter 9, NASA helped pioneer the implementation of image-processing techniques to enhance and correct pictures transmitted by outer-space probes. Similar techniques have been applied on Earth in applications ranging from desktop publishing to those in the medical field. Remote-sensing satellites, originally designed to explore the Earth, have also been used by the news media. Consequently, there is a cross-fertilization between what can be called outer- and inner-space operations, and as such, outer-space developments merit, at the very least, an overview.

This section also provides a brief history of NASA and, more pointedly, highlights some of the forces that have shaped and continue to shape the space program, including social and political issues and pressures. The section concludes with an overview of legal implications governing space-based activities.

History

The 1950s witnessed the birth of the modern era of space exploration.[38] A milestone in this process was the inauguration of NASA on October 1, 1958, as the successor to the National Advisory Committee for Aeronautics (NACA). The latter, founded in 1915, had played a key role in the development of the nation's aeronautical industry through research and related activities. The new agency had a similar mandate and was also delegated the task of overseeing the civilian space program.

Some of the major events in the space program, as well as influencing factors, include the following:

1. The Soviet Union launches the first artificial satellite, Sputnik 1, on October 4, 1957.

2. On April 12, 1961, Soviet cosmonaut Yuri Gagarin becomes the first human in space. The U.S. space program centers about Project Mercury and its seven astronauts.

3. President John F. Kennedy commits the nation to landing an astronaut on the moon before the end of the decade (Project Apollo).

4. On January 27, 1967, a fire in the *Apollo* command module kills three astronauts.

5. On July 20, 1969, Neil Armstrong and Edwin ("Buzz") Aldrin of *Apollo 11* become the first humans to walk on the moon, while

their comrade, Michael Collins, orbits overhead.

6. The Apollo program comes to a halt after *Apollo 17* in December 1972, owing to financial considerations and the changing U.S. social and political climate (such as the Vietnam War).

7. The 1970s and early 1980s witness other missions, including the Apollo–Soyuz Test Project (1975), where a Soviet and U.S. spacecraft dock in orbit, and the development of the space shuttle.

8. Satellites and space probes explore the Earth and most of the planets in the solar system. For example, Landsat remote-sensing satellites image the Earth for agricultural, geological, and more recently, for news-related applications.

9. Budgetary constraints and the loss of *Challenger* contribute to the dearth of planetary missions from the late 1970s until the late 1980s. The potential lack of appropriations for projects, such as the ACTS satellite, could have similarly derailed the development of enhanced satellite communications technologies.

10. The Galileo probe will investigate Jupiter and its moons in the 1990s while the Hubble Space Telescope continues its exploration of the universe from a low Earth orbit. Both systems, though, have suffered from performance problems.

Legal Implications

The capability to use and explore both inner and outer space has legal implications. As our satellites' capabilities increase, and as we develop commercial space-based enterprises, the national as well as international bodies of law that may govern these activities become more important. Relevant topics already discussed include orbital assignments, signal piracy, and the role of Intelsat and competing organizations in the international satellite communications arena.

Remote sensing, DBS, the national sovereignty of airspace, and individual rights are four additional subject areas. The United Nations has, for instance, been concerned with the international free flow of information and the cooperation between nations in the

dissemination of the data generated by remote-sensing satellites. In a related area and on a national level, the use of remote-sensing images by the news media has been, as discussed, the target of government restrictions and regulation.

Questions were raised about the socio-political impact of DBS relays if the signals were picked up by a country other than the host nation.[39] Concerns were also expressed about preserving the free flow of information and ensuring that it was balanced: the information stream would not flow solely from the developed to the developing world.

The third topic: how high is high? Basically, what is the "upward extent" of a country's national sovereignty? Is it 100 miles, or could it be lower or even higher?[40] In a complementary fashion, what are the political implications for satellites with respect to their orbital slots?

The final question, at least for our discussion, is that of individual rights for space explorers and/or colonists. The U.S. Constitution Bicentennial Committee covered this issue in a subcommittee. An outcome, the *Declaration of First Principles for the Governance of Outer Space Societies*, declared that the U.S. Constitution should also apply to "individuals living in outer space societies under United States jurisdiction."[41] The document's drafters believed that individual rights, such as freedom of speech, assembly, and *media and communications* (my emphasis), are fun-

Figure 6–9
Satellite systems have enhanced our communications capabilities and our ability to cover world events. The impact of the Kuwait oil fires (for example, smoke), set during Operation Desert Storm, is evident in this sequence of photos. (Courtesy of the Earth Observation Satellite Company, Lanham, Md.)

KUWAIT
OIL FIRES

January 6,
February 15,
& March 3, 1991
Bands 4,3,2 (RGB)
Path/Row 165/40

EOSAT
Earth Observation Satellite Co.
Lanham, Maryland, USA

damental principles that would extend to U.S. space societies, balanced against the unique environment afforded by outer space.[42]

Finally, there are other legal issues beyond the scope of this book, such as licensing policies and procedures for communications satellites.[43] For the realm of outer space, there is a growing body of space law. It is a fascinating field, and one that will continue to evolve as we begin to take our first outward steps in space.

REFERENCES/NOTES

1. *Telecommunications: A Glossary of Telecommunications Terms* (Federal Communications Commission, April 1987), 4, 14.
2. Intelsat, "Intelsat VI (F-3); Reboost Mission," press packet, 4.
3. Mark Long, *World Satellite Almanac* (Boise, Idaho: Comm Tek Publishing Company, 1985): 73.
4. *Stabilization* refers to the manner in which a satellite maintains its stability while in orbit. A spin-stabilized satellite rapidly rotates around an axis while the antenna is situated on a despun platform so it continues to point at the Earth. Three-axis-stabilized spacecraft use gyros to maintain their positions. Other systems also play a role in this process. NASA has used a three-axis-stabilized design for many of its outer-space probes. Please see Andrew F. Inglis, *Satellite Technology* (Boston: Focal Press, 1991), 32; and P. R. K. Chetty, *Satellite Technology and Its Implications* (Blue Ridge Summit, Penn.: TAB Books, 1991), 174–79, for more detailed information about satellite stabilization systems.
5. Besides activating the thrusters, another vital element in the transmission between a satellite and a ground station is telemetry data. The telemetry information stream essentially is housekeeping data relayed by the satellite to indicate its current operational status. In addition, the ground station uplinks commands to the satellite to initiate a variety of operations, such as the aforementioned activation of the thrusters. In some instances, a command may be issued to move the satellite to a completely different orbit. Thus, depending upon its design, a spacecraft is not necessarily locked into a position once it achieves its assigned slot. It may be able to move to different positions.
6. Comsat, "Comsat Maneuver to Add Life and Versatility to Satellites," press release #86–35, October 16, 1986, 3.
7. A controversy erupted about RTGs prior to the launching of the Galileo spacecraft. Galileo is designed to investigate Jupiter and its moons during an extended space mission. Its systems are powered by RTGs, and protesters legally sought to block its launch due to a concern over the possible contamination of the Earth if the RTGs' fuel was scattered in a launch disaster.

   NASA and space advocates replied that the nuclear fuel was placed in protective containers. Even if there was an explosion, the fuel would not be scattered. Ultimately, the spacecraft was launched in October 1989. Robert Nichols provides an excellent overview of this issue in his article "Showdown at Pad 39-B," *Ad Astra* 1 (November 1989): 8–15.
8. Conversation with Scott Bergstrom, Ph.D., director, Technology-Based Instruction Research Lab, Center for Aerospace Sciences, University of North Dakota, August 6, 1992. For additional information, see Peter Lambert, "Digital Compression; Now Arriving on the Fast Track," *Broadcasting*, July 27, 1992, 40–46.
9. Satellite Communication Research, *Satellite Earth Station Use in Business and Education* (Tulsa, Okla.: Satellite Communication Research), 18. Also see Andrew F. Inglis, *Behind the Tube* (Boston: Focal Press, 1990), 405, for additional information.
10. Mark Long, *World Satellite Almanac* (Boise, Idaho: Comm Tek Publishing Company, 1985), 99.
11. Gerhard Hanneman, "Current Status of Teleports in the Americas: 56 Operational, Planned, or Under Development," *Communications News*, March 1986, 43.
12. John Pavlik and Mark Thalhimer, "The Charge of the E-Mail Brigade: News Technology Comes of Age," in *The Media at War: The Press and the Persian Gulf Conflict* (New York: Gannett Foundation Media Center, 1991), 35.
13. "Group Launches Campaign to 'Pull Plug' on CNN's Arnett," *Broadcasting*, February 18, 1991, 61.
14. Pavlik and Thalhimer, "Charge of the E-Mail Brigade," 37.
15. Jay Peterzell, "Eye in the Sky," *Columbia Journalism Review*, September/October 1987, 46.

16. In one case, a 10-meter resolution limitation was lifted. It should also be noted that other countries have gotten on the remote-sensing bandwagon, including Russia for very high resolution images. Please see the "Land Remote Sensing Policy Act of 1992," S. 2297, 102d Congress, 2d session, for specific information about U.S. policy issues.

17. HBO provided the impetus for the development of the VideoCipher system, originally developed by M/A-Com, Inc. This concept was described in a 1983 HBO brochure, "Satellite Security," for its affiliates.

18. William Sheets and Rudolf Graf, "The Raid on HBO," *Radio-Electronics* 10 (October 1986): 49.

19. Gary M. Hoffman, *Curbing International Piracy of Intellectual Property* (Washington, D.C.: The Annenberg Washington Program, 1989), 11.

20. Peter Lambert, "Countdown to Renewable Security," *Broadcasting*, July 27, 1992, 56.

21. Hughes Communications, Inc. "Staying Clean," *Uplink*, Spring 1992, 6.

22. "Commerce Department Sees Bright Future for Advertising," *Broadcasting*, January 12, 1987, 70.

23. "USSB, Hughes Revive DBS in $100 Million+ Deal," *Broadcasting*, June 10, 1991, 36.

24. Hughes Communications, Inc., "DirecTV," information flyer.

25. Ronald J. Schertler, "ACTS Experiments Program," *A NASA Technical Memorandum prepared for Globecom '86, sponsored by the Institute of Electrical and Electronic Engineering*, Houston, Tex., December 1–4, 1986, 3.

26. Following this mission, the satellites were subsequently recovered and returned to Earth via a shuttle.

27. William P. Rogers, Neil Armstrong, David C. Acheson, et al., *Report of the Presidential Commission on the Space Shuttle* Challenger *Accident* (Washington, D.C.: U.S. Government Printing Office, 1986), 201.

28. Yale Jay Lubkin, "What Really Happened," *Defense Science* 9 (October/November 1990): 10.

29. The Reagan administration had been a proponent of the commercialization of outer space, especially in the area of the launch industry. Please see Edward Ridley Finch, Jr., and Amanda Lee Moore, *AstroBusiness* (Stamford, Conn.: Walden Book Company, 1984), 56–63, for more information.

30. Brian J. Horais, "Small Satellites Prove Capable for Low-Cost Imaging," *Laser Focus World* 27 (September 1991): 148.

31. As of this writing, a Pegasus launch would cost approximately $7 to 10 million versus millions of more dollars for a standard ELV.

32. It should be noted that salvage missions of this nature are limited, at least at this time, to low-Earth-orbit operations.

33. The ESA and its member states support a wide range of space activities. These have included developmental work with ELVs, space exploration via outer-space probes, and the development of a shuttle-like vehicle. Please see the special advertising supplement to *NASA Tech Briefs* 15 (June 1991) and *Ad Astra* 3 (December 1991): 18–40, for more specific information about the ESA.

34. Dr. R. T. Gedney, "Foreign Competition in Communications Satellites Is Real," *ACTS Quarterly*, Lewis Research Center, issue 91/1, February 1992, 1.

35. U.S. General Accounting Office, *National Aero-Space Plane; A Technology Development and Demonstration Program to Build the X-30*, GAO/NSIAD-88-122, April 1988, 14. Please note: The X-30 is also a hypersonic flight vehicle. As stated on p. 10 of the same report, "Hypersonic is that speed which is five times or more the speed of sound in air (761.5 mph at sea level)." The X-30 could reach hypersonic speeds up to Mach 25.

36. U.S. General Accounting Office, *National Aero-Space Plane*, 50.

37. See Jim Martin, "Creating the Platform of the Future; NASP," *Defense Science*, September 1988, 55, 57, 60, for more information about the NASP program, including potential military applications. NASP-based vehicles would not eliminate ELVs or possibly even the shuttle. This is one lesson we've hopefully learned from the shuttle program: one vehicle may not be able to perform all functions equally well.

38. Parts of this section are taken from Michael Mirabito, "Space Program," in *The Reader's Companion to American History* (New York: Houghton Mifflin Company, 1991), 1013–14. Houghton Mifflin kindly extended permission for its use.

39. Stephen Gorove, "The 1980 Session of the U.N. Committee of the Peaceful Uses of Outer Space: Highlights of Positions on Outstanding Legal Issues," *Journal of Space Law* 8 (Spring/Fall 1980): 179.

40. S. Houston Lay and Howard J. Taubenfeld, *The Law Relating to Activities of Man in Space* (Chicago: The University of Chicago Press, 1970), 41.

41. Nathan C. Goldman, "Space Colonies: Rights

in Space, Obligations to Earth," in Jill Steele Meyer, ed., *Proceedings of the Seventh Annual International Space Development Conference* (San Diego: Univelt, Inc., 1991), 220.

42. Rights versus the space environment include the right to bear arms, an important principle in the United States. But in space, where a weapon could physically compromise the integrity of a colony's protective shielding, this issue becomes more complex. The same question applies to the freedom of assembly and the press, among others. This general concept has also been used as the plot in various works, including Robert A. Heinlein's science-fiction book *The Moon Is a Harsh Mistress* (New York: Berkeley Publishing Corp., 1968). Please see William F. Wu, "Taking Liberties in Space," *Ad Astra* 3 (November 1991): 36, for more information.

43. Please see Carl J. Cangelosi, "Satellites: Regulatory Summary," in Andrew F. Inglis, ed.-in-chief, *Electronic Communications Handbook* (New York: McGraw-Hill Book Company, 1988), 6.1–6.9, for additional information.

ADDITIONAL READINGS

Abutaha, Ali F. "The Space Shuttle: A Basic Problem." This videotape was produced by George Washington University, Washington, D.C. It is a taped lecture from a course conducted by Ali Abutaha for George Washington's Continuing Engineering Education program. The tape covers some of the design problems of the shuttle discovered by Abutaha after the *Challenger* disaster.

Brender, Mark E. "Remote Sensing and the First Amendment." *Space Policy.* November 1987, 293–97. An excellent review of the remote-sensing field in regard to journalism and First Amendment issues. Also details news applications of the photographs.

Chetty, P. R. K. *Satellite Technology and Its Applications.* Blue Ridge Summit, Penn.: TAB Professional and Reference Books, 1991. Examines satellites and their applications. The book is especially strong in its coverage of satellite design.

Communications News. From late 1987 through 1988, *Communications News* ran a series of articles devoted to VSATs. The series ran under the main title "VSAT Technology for Today and the Future," with each article having its own subtitle. An example is David Wilkerson, "VSAT Technology for Today and the Future—Part 3:

Use Private Networks or Leased Services?" November 1987, 60–63.

Dorr, Les, Jr. "PanAmSat Takes on a Giant." *Space World* W–12–276 (December 1986): 14–17. And, Anselmo, "Landman Team Up to Tackle Intelsat." *Broadcasting,* August 5, 1991, 48. Both articles cover the development of private, international satellite networks.

EOSAT. "Special Landsat 6 Issue." *Landsat Data Users Notes* 7 (Summer 1992). A look at the next generation of Landsat satellites, Landsat 6.

"EOSAT Operations Underway." *The Photogrammetric Coyote* 9 (March 1986): 8, 17. "SPOT to Fly in October." *The Photogrammetric Coyote* 8 (September 1985): 2, 4. A brief introduction to the Landsat and the SPOT satellites, respectively, as well as their controlling organizations.

Finch, Edward Ridley, Jr., and Amanda Lee Moore. *AstroBusiness.* Stamford, Conn.: Walden Book Company, 1984. An excellent primer to the various issues surrounding the commercialization of outer space, including the development of new launch vehicles and relevant space law as well as space-insurance issues.

Hannigan, Russell J. "Europe Eyes the 'Bush Push.'" *Ad Astra* 3 (February 1990): 9–13. An examination of some of the issues concerning U.S. and European cooperation in space ventures.

"Hughes's Petrucci and Hartstein: Taking the Direct Approach." *Broadcasting.* July 29, 1991, 52–55. Interview with two Hughes Communications executives and the status of DBS systems.

Inglis, Andrew F. *Satellite Technology: An Introduction.* Boston: Focal Press, 1991. An excellent guide to satellite technology and applications.

Kerrod, Robin. *The Illustrated History of NASA: An Anniversary Edition.* New York: Gallery Books, 1988. A comprehensive and richly illustrated history of NASA and the U.S. program.

Khan, Tariq. "'Third Generation' Technology Fuels VSAT Growth." *Telecommunications.* September 1990, 29–30, 32, 34. An overview of the evolution of VSAT technology.

Lewis Research Center, NASA. *ACTS Quarterly.* A newsletter published by the Lewis Research Center that traces the development of the ACTS satellite. As a whole, the collection provides an interesting overview of the growth of a satellite program and the impact of external forces, such as budgetary appropriations.

Niekamp, Raymond A. "Satellite Newsgathering and Its Effect on Network-Affiliate Relations." An AEJMC Convention paper, Radio-TV Journalism Division, August 1990. A study of network affiliates and "their policies in sharing news video with their networks and other stations."

U.S. General Accounting Office. *Aerospace Plane Technology: Research and Development Efforts in Japan and Australia.* GAO/NSIAD–92–5, October 1991. One of a series of reports that provides an international look at aerospace plane technology, applications, and developments.

GLOSSARY

Active satellite: A satellite equipped to receive signals and to relay its own signal back to Earth.

Advanced Communications Technology Satellite (ACTS): A NASA satellite that is a prototype of the future commercial satellite communications fleet.

C-band: A satellite communications frequency band and a satellite class that uses this band for communication purposes. Commercial C-band satellites are the older and more established of the contemporary communications satellite fleet.

Communications Satellite Corporation (Comsat): The U.S. representative to Intelsat.

Direct-broadcast satellite (DBS): As originally conceived, a powerful communications satellite that would deliver movies and other offerings to subscribers equipped with compact satellite dishes.

Earth station: One member of the Earth-based segment of a satellite communications system. The Earth station comprises an antenna or dish and a transmitter that can relay a high-frequency microwave signal. Some Earth stations, also called ground stations, can transmit and receive signals, while other stations are receive-only configurations.

Footprint: The shape of a satellite's transmission in regard to the reception area on the Earth.

Geostationary orbit: A desirable orbital position or slot for a communications satellite. When placed in such an orbit, the satellite's motion is synchronized with the Earth's rotation and appears, at least to a ground observer, to be stationary in the sky. Prior to this time, a ground station had to track a satellite as it moved across the sky. This made ground activity more complicated.

International Maritime Satellite Organization (Inmarsat): An international organization that extended satellite-based communication to ships at sea, oil-drilling platforms, and even remote sites on land.

International Telecommunications Satellite Organization (Intelsat): An international satellite consortium. Intelsat supports a broad range of satellite services.

Ku-band: A satellite communication band that has been more recently adopted. Ku-band satellites, for example, can generate a more powerful downlink than their C-band counterparts.

Launch vehicles: Both expendable launch vehicles (ELVs) and vehicles that carry passengers (for example, the space shuttle) have been used to launch satellites. Organizations involved in this venture range from NASA to Arianespace.

Lightsat: A small, relatively inexpensive satellite.

Mediasat: A proposed remote-sensing satellite designed to take high-resolution images of the Earth for media news applications.

Orbital spacing: Satellites are positioned in the orbital arc above the Earth, and buffer zones physically separate the satellites to help eliminate any potential interference.

Passive satellite: A satellite that does not carry the necessary equipment to relay its own signal back to Earth.

Remote sensing: The process by which a special class of satellite, a remote-sensing satellite, scans and explores the Earth. Through a variety of instruments, including special cameras, images of the Earth are produced that can highlight its physical characteristics (for example, the loss of wetland acreage). The media also have a stake in this field, since photographs of different regions of the Earth that are normally inaccessible to news cameras can be produced by remote-sensing satellites.

Satellite newsgathering (SNG): The term

that describes the use of small, transportable satellite dishes to directly relay news stories from almost anywhere in the field (via satellite).

Scrambling: The technical process by which a satellite's signal is rendered unintelligible. The receiving site is equipped with a special decoder to return the signal to its original state.

Teleport: A satellite dish farm.

Transponder: The heart of a satellite's communications system that acts like a repeater in the sky.

Very Small Aperture Terminal (VSAT): A VSAT configuration is a small satellite dish and the complementary electronic components. A VSAT system is relatively inexpensive to establish, and a dish may be capable of operating in a receive-only or in a receive-and-transmit mode.

7

Desktop Publishing

INTRODUCTION

This chapter examines *desktop publishing* (DTP), an application that cuts across the computer and communications fields. Desktop-publishing systems are used by large and small organizations, and by individuals, to produce newsletters, brochures, magazines, and other documents.

The tools for this application include PCs, software, printers, and peripherals such as *scanners*. When combined, they create a near-typeset-quality publishing system that can literally fit on a desktop.

This same system also provides an individual with an electronic composition tool that can make formerly inaccessible publishing and printing capabilities practical. For example, instead of following the "cut-and-paste"

method to create a layout, where text and graphics are manually cut and positioned on a page, a design is electronically implemented through a computer and DTP software. A monitor's screen serves as a window in this operation and depicts the size and location of text, headlines, and graphics.

A combination of factors led to the proliferation of PC-based publishing systems in the mid- to late 1980s. Sophisticated PCs flooded the market while DTP and other software packages that took advantage of the computer's processing capabilities were developed. The final element in this configuration, the laser printer, was likewise introduced at an affordable price.

It is important, at this point, to view the DTP outfit as a system, since the various components must work together. A program

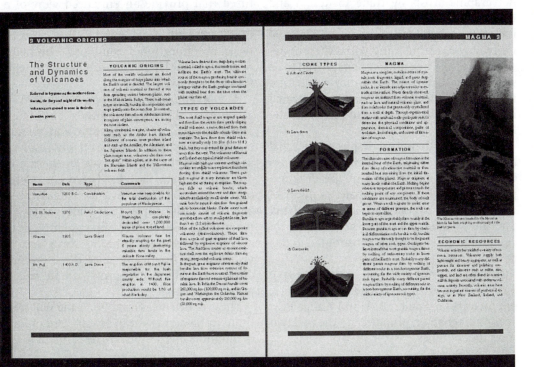

Figure 7–1
DTP software allows you to create complex documents that can combine text, graphics, and information tables, among other elements. (Reprinted with permission of and copyright © 1986–1992 Frame Technology Corp. All rights reserved; FrameMaker)

should be able to tap the printer's various features while the printer, in turn, should be equipped to support the job at hand and, if possible, anticipated jobs.

HARDWARE

The Computer

The computer that helped launch the PC-based DTP industry was the Macintosh. The Macintosh was developed as a graphics-oriented machine; moreover, the Macintosh hardware was matched by a complementary selection of software that tapped the computer's capabilities.

The Macintosh was also designed as part of an integrated hardware and software system. In contrast to some IBM configurations, it was generally easier to get a Macintosh configuration up and running. A stock Macintosh system was also better equipped to handle desktop-publishing and graphics applications than its IBM counterpart, and in the eyes of some computer owners, the Macintosh's GUI made it easier to use.

When combined, these factors made it possible for a new group of users who were not graphic artists by trade to design and complete their own creations. Graphic artists, on the other hand, now could own a tool for readily experimenting with different design concepts. The entire system proved to be so popular that rival computer companies manufactured machines that incorporated many of the features popularized by the Macintosh.

IBM PCs, for their part, despite some initial hardware and software disadvantages, also emerged as a major force in the desktop-publishing field. The dominant position of IBM PCs in the overall computer market and the introduction of new generations of equipment and programs contributed to this development.

Other platforms and systems support DTP operations. But as described in Chapter 3, our focus is on Apple, IBM, and Amiga PCs, and in this chapter, primarily on the first two computer families.

The Monitor

The second major component of a DTP system is the monitor; either a standard color or monochrome unit can be used. But the typical 14-inch or smaller monitor can't display a full page of readable text, and different viewing modes must be employed during the design process.

In the full-page mode, a page can be displayed in its entirety as it is composed. This option is valuable when a page is initially designed, since it reveals the placement and the spatial relationship between the graphic and textual elements.

The page outline is noticeably reduced in size when this mode is selected. Most if not all of the text may be replaced by small lines or bars (a process called *greeking*) since the characters are essentially too small to be reproduced on the screen.

The other modes provide magnified or enlarged views of specific sections of a page. In this way, the text can be read and fine details of the document's style can be checked. Thus, during an actual operation, both full-page and magnified views are used as the work progresses.

Special monitors have also been manufactured that provide an enhanced view of a document, especially when the software is used in the full-page mode. Larger, conventional monitors, when combined with high-resolution display modes, are likewise helpful in this regard.

A conventional monitor's resolution is also not equal to a laser printer's capabilities, and consequently, the graphic and textual elements may not appear as finely resolved as they do in the printed copy. Furthermore, due to software and hardware considerations, earlier DTP systems may not have been able to clearly display the wide variety of typestyles a laser printer could produce. Both characteristics made it harder for a user to visualize the final copy since the monitor's display could only approximate a page's printed appearance.

These deficiencies have been and will continue to be addressed in future generations of monitors and software. The result of these developments will be a display that more closely matches the final copy, and a user will be able to exercise a greater degree of control over a document as it is created.

Printers

The third component of a PC-based DTP system is a laser printer, a device capable of producing documents that are near-typeset in quality. Most laser printers can create documents with a 300-dots-per-inch (dpi) resolution. In the context of our discussion of printers, the term *resolution* refers to the apparent visual sharpness or clarity of the printed characters and graphics as they appear on the page. This working definition is used throughout the chapter.

There is a relationship between the dpi figure and the print quality of a document: as this figure increases, so does the number of dots that make up the alphanumeric and graphics information. The higher dpi figure would ultimately result in a document with a superior physical appearance, much like a document that has been printed by one of the more sophisticated and powerful commercial units.

When the first reasonably priced laser printer appeared in 1984, it created a stir in the computer industry. Since the printer could produce high-resolution documents in a wide range of typestyles, it could accommodate various printing jobs that had normally been reserved for traditional typesetting equipment. This trend, started by the Hewlett-Packard LaserJet and the Apple LaserWriter printers, has continued unabated.

As of this writing, most if not all printers designed for the general business and consumer markets share several broad characteristics.

First, a printer should be equipped with a megabyte or more of memory to take full advantage of its printing capabilities. This requirement increases for a color printer.

Second, there is a diversity of *typefaces* and *fonts* available for most printers. This factor enables you to design a document that fits your specific requirements.

A typeface is a specific and unique print

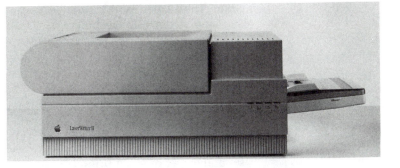

Figure 7–2
The Apple LaserWriter II. (Courtesy of Apple Computer, Inc.) [photo credit: John Greenleigh]

style. The different characters of a given typeface conform to a style, a set of physical attributes shared by all the characters. Two examples of common typefaces are Helvetica and Times Roman. A font, in contrast, is a typeface in a specific size. The size is measured in points, and as the point size increases, so too does the character's size. As a frame of reference, 72 points approximately equals one inch. Laser printers can support numerous typefaces and fonts, which can be mixed and printed on the same page.

A printer is equipped with a number of fonts, the font library, when it is initially purchased. Additional libraries are available on disks, from which they are subsequently downloaded or relayed to the printer by the computer. The fonts may also be stored on a special, removable cartridge that attaches to the printer or they may be stored on a dedicated hard drive.[1]

Third, although the standard 300-dpi laser printer cannot generate typeset-quality documents, it is satisfactory for creating newsletters, an organization's in-house magazine, or when applicable, books on a tight production schedule and budget. High-quality line drawings, a series of black lines on a white background, can also be printed. These may include a building, a chart, or the interior view of a jet engine that will be used in a technical document.

The standard laser printer also continues to be improved. Over the next few years, printers with an enhanced gray-level capability and at least twice the resolution of current models should dominate the general market. Third-party products are also available that perform a similar function.

Fourth, the typical laser printer cannot support a full-color output. Its graphics capabilities are also limited for reproducing black-and-white photographs. Details may not be sharply defined or too few gray levels may be reproduced.

Color Printing. As indicated, the typical laser and even non–laser printer is not color ready. This situation is gradually changing, though.

In the early 1990s, color printers became more popular as the price for these units started to fall. This includes a series of color printers that employ a thermal-wax transfer method. In this system, printing takes place by "heating colored wax (cyan, magenta, yellow, and black) and fusing it to a special paper. The print head melts tiny dots of color from a ribbon onto the paper."[2] Thermal-wax printers, as of this writing, could cost less than $5,000 per unit.

Color printers suitable for general business and consumer applications also include low-end ink-jet and dot-matrix printers. Both printer types are cost-effective but typically can't match a thermal-wax printer in quality. Laser color printers also exist. They produce a superior output but are still comparatively expensive. Other color printers, such as dye-sublimation units, are also available.[3]

Depending on your needs, you may even opt for commercial printing. While new hardware and software releases may make it easier to prepare color images for this task, it can still be a complicated process.[4] Consequently, it's a good idea to discuss a project with a commercial printer to obtain the best results. You may also decide to leave the work to individuals who are well versed in color theory and operations.[5]

Finally, color can be an important element of a DTP project, be it produced by a commercial printer or by a machine in your office or home. Color can catch a reader's eye and, when used properly, can help convey information more effectively.

Think of a bar chart showing one radio station's ratings compared with the other stations in the market. The use of different colors makes it easier to differentiate between the chart's bars and, thus, the information itself. In another example, the use of color may produce a more visually appealing ad.

PostScript. The way a page is actually printed can be illustrated by *PostScript,* a popular printing method. PostScript is a page-description language (PDL). It was developed by Adobe Systems, Inc., popularized by Apple, and has emerged as a standard in the DTP field.

In essence, PostScript is one of the software mechanisms that makes it possible for a laser printer to reproduce or print, in a hard-copy form, the information generated by a DTP program. PostScript is also a programming language

whose sole purpose is to precisely describe the placement and appearance of text and graphics on a page or pages. . . . [T]he language must establish conventions for the printing device. These include telling it . . . what type fonts are in use. . . . PostScript is an interpreted language. . . . [T]he computer must "interpret" the statements into a machine-executable form. A PostScript printer must have its own internal computer to do this.[6]

Thus, PostScript defines the graphics and alphanumeric elements on a page. It is the interface between the DTP program and the laser printer.

PostScript is a very flexible standard. During the printing process, a letter such as *A* would initially be defined. Next, when a different point size is selected, the new letter is automatically computed based upon the original description and information. This means a library of the letter *A,* in a variety of sizes, does not have to be stored and retrieved when a new size is selected.

A PostScript-based system is also hardware independent, unlike some other standards in the field. This is a valuable feature since a document can be manipulated by a wide range of equipment, and you are not tied or locked into one hardware family. A typical scenario is as follows.

A document can be composed and saved with a DTP program that supports a Post-

Script driver. In our example, the document is initially generated on a 300-dpi laser printer to create a proof copy. The same file can then be used with a 1,000-plus-dpi PostScript-compatible, professional printing system, to produce a superior final product. Consequently, a document's highest resolution is not set at 300 dpi or any other resolution, and the quality of the copy is limited only by the equipment that is used.

It is the PostScript standard, in this instance, that makes it possible to tap a printing device's highest resolution mode. This characteristic contributes to PostScript's forward and backward compatibility. As newer and higher-resolution devices are introduced, they can be used at their higher-resolution settings. Older printers and other devices can also be accessed even though the document would not be as refined as those produced by the higher-dpi units.

Some printers already include PostScript when sold; others can be fitted with a PostScript cartridge. For a less-expensive option, you can use a special program, a PostScript interpreter, that taps this standard's capabilities without additional hardware. As an added bonus, the software supports dot-matrix and ink-jet printers, so their owners can also take advantage of a PostScript-based configuration.

Finally, there is a price for using this standard. A PostScript printer is more expensive than a comparable non-PostScript model. Printing may also be slow, depending upon the complexity of the information, the printer, and other factors.[7] Recent enhancements to the PostScript language have, however, improved the printing time as well as other operations.

SCANNERS

Besides using conventional graphics produced by paint and drawing programs, a DTP project may include black-and-white photographs, line drawings, and other artwork originally produced in a hard-copy form. This operation is made possible by using a scanner, a piece of equipment that is interfaced with the computer.

In one design, a photograph can be placed on the scanner, much like a piece of paper on a copy machine. The image is then read or "scanned" by the device, and the picture information is digitized and fed to the computer.

When the photograph is actually reproduced on a laser printer, a digital *halftone* process can be employed. A conventional laser printer cannot print true shades of gray and only prints black dots on a page. Consequently, a halftone method is used to create gray-level representations or simulated gray shades on the final printed copy.

In this process, the image is divided into small areas or cells. The various gray-shade representations in the picture are subsequently generated by turning dots in these small areas either on or off. This varying density of black dots, and ultimately the cells, creates the various apparent shades of gray throughout the picture.

Other factors play a role in this operation. But the result is a picture that generally can't match the quality of pictures reproduced in magazines and similar publications, which use an analogous but more sophisticated version of this process.

There is also a balance, especially when using a typical PC-based configuration, between the number of gray levels and the picture's resolution. As the number of levels increases, the apparent resolution drops.[8] Line drawings, however, are not affected by this factor since there are no intermediate gray shades that must be reproduced.

After a picture is scanned, it can also be saved in a format that is compatible with graphics and image-editing programs. This capability is the basis for a powerful software tool. As described in Chapter 3, an image can be edited, the contrast and brightness levels can be changed, and special filters can be applied to further manipulate the image. At this point, the new version of the original image is saved and imported by the DTP software.

Gray-Level and Color Work. When scanners were initially introduced in the general market, they were not able to capture gray-level information. This has changed, and the

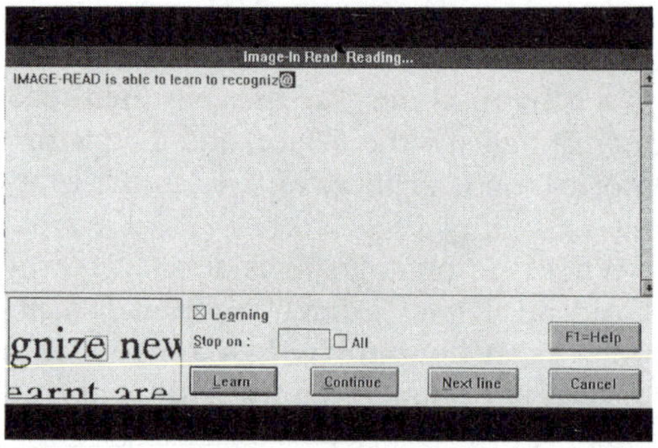

Figure 7–3

An OCR operation. The text is reproduced in the top of the screen (top window). The window in the bottom left corner shows the original scanned text. The learning mode is also activated so the system can be "taught" to identify letters, for example, that were previously unrecognized. (Software courtesy of Image-In, Inc.; Image-In-Read)

typical scanner can now support 256 levels of gray. The upshot of this capability is an enhanced photographic-reproduction quality, especially when using one of the professional printers. Capturing gray-scale information also allows you to take full advantage of image-editing software.[9]

In addition to gray-scale images, a scanner may support color. Different scanning approaches are used to capture the color information. In one system, the image is scanned three times, using different filters, and the final product is an RGB composite image.[10] Depending upon the scanner, a 24-bit-color mode may also be supported.

This ability to scan preexisting pictures and graphics is an important one, for it supports applications beyond the desktop-publishing field. As discussed in the next chapter, scanned images may be a central element of a desktop video or a multimedia project. Gray-scale and color scanners, when combined with image-editing software, can form a crucial link in the production process.

Optical Character Recognition. In addition to capturing images, a scanner can be used with *optical character recognition* (OCR) software. Scan a printed document, and the software recognizes the alphanumeric information. The information can then be saved as a file and retrieved with a word-processing program.

The advantage of an OCR system is its labor-saving feature. Instead of retyping different documents, which is a time-consuming and tedious job, you can let the scanner do the work. Next, a word-processing program can be used to edit the information as

easily as you would edit a document that was originally typed on the computer.

But there are certain limitations with an OCR operation. The software may recognize only a limited number of typestyles, and the characters must be legible and fairly dark for this operation. Some characters will also be incorrectly read and must be replaced during an editing session.

To overcome one of these problems, the capability to read a limited number of typestyles, some systems support a learning mode. If a document is printed in an incompatible format, you can teach the computer to recognize this new information. More recent OCR software can also recognize a wider range of type and different point sizes.

Other Developments. As indicated, a scanner can be a very valuable addition to a DTP system. Preexisting artwork and photographs can be integrated in a document, and when combined with the proper software, an OCR operation can be supported.

It is also possible to use cost-effective hand scanners. These portable units, as implied by the name, can be held in the hand and have made scanning available to a broader user base. You physically move the scanner down the image or page of text to capture the information.

While hand scanners can support gray levels and color, they do have one major disadvantage. Since their width is limited, you may have to complete two hand scans to cover an entire page. Nevertheless, special software can join or stitch the two scans together to produce the full image. The quality of this operation can vary and depends upon the software and how straight the original work is scanned.

Newer controlling programs have also improved a scanning operation for many users. Led by Ofoto, a package originally released for Apple-based systems, higher-quality images can be automatically generated. Before this development, it may have taken a longer time to produce an acceptable scan.

This capability also serves an important function. The interface simplifies the scanning process, which makes the technology more accessible. Essentially, you don't have to be an expert in this area. You supply the aesthetic framework, and the program can help you produce a better product. Manual adjustments and various image-editing operations may also be supported.

In a related area, film scanners are becoming more popular for specific tasks. Newspapers, for instance, have used such systems to send photographs between local, national, and international locations.

In a typical situation, a roll of film is developed, and select film images are digitized by a scanner. They are subsequently compressed, relayed over a telephone line, and received and reproduced at the home office.[11] This system saves time and has provided for a more error-free relay.

THE SOFTWARE

The heart of any DTP system is the software. Our discussion focuses on two PC program categories that are the most popular at this time, word-processing and page-composition programs.[12]

Word-Processing Software

Sophisticated word-processing programs can be used to complete some jobs normally reserved for DTP software. These include generating articles, business forms, newsletters, and other publications.

A program incorporates an assortment of features to assist the writer, including a search function that quickly pinpoints a specific word in the entire text file, and more specialized aids, such as macros. A macro is a user-defined command that performs a specific task with only one or two rather than with multiple keystrokes.

Some word-processing programs are now so comprehensive that they are more properly categorized as a cross between a traditional word-processing and a DTP program. For example, a program may produce a *What You See Is What You Get* (WYSIWYG) display where a document's screen appearance approximates the final printed copy. This feature simplifies the design process since you can see all the elements that compose the document as it is created.

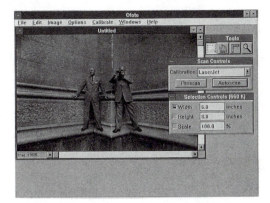

Figure 7–4
Some of the options of the Ofoto software package. (Courtesy of Light Source)[photo credit: © Rick Smolan]

A word-processing program may also support graphics on the same page as text. Instead of placing an illustration on a separate page, it is incorporated in the relevant body of text, much like a newspaper or magazine. Thus, a graphic that depicts the rise and fall of the stock market over the last decade can be inserted on a page of text that describes this situation. Other advanced features may include the capability to create

- multiple columns;
- simple graphics;
- tables and charts;
- a table of contents and an index; and/or
- scientific and mathematical equations.

In sum, a sophisticated word-processing program can be very useful in a wide range of applications, including those that do not demand the full power of a DTP system. A word-processing program is also typically easier to use, faster, and like a DTP program, can produce near-typeset-quality documents when combined with a laser printer.

Page-Composition Software

Page-composition programs, which may be

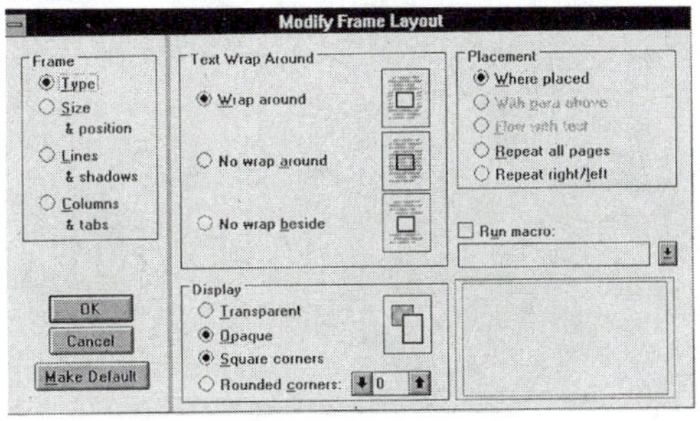

Figure 7–5
Newer word-processing programs also support enhanced design and formatting capabilities, such as an option to create frames that can, for example, incorporate images or text. (Created with Ami Pro. Ami Pro is a registered trademark of Lotus Development Corp. © 1993 Lotus Development Corporation. Screen dump used with permission.)

considered "true" DTP software, support an interactive interface that generates a WYSIWYG screen display. As you change the position of a graphic or a column of text, the change takes place in real time, or as you actually move the graphic or text to a new position. This visual feedback helps you determine if a design for a page is satisfactory, and it allows you to experiment quickly with different designs.

The importance of this interactive system, where you literally interact with the computer and the software to implement a design, cannot be overemphasized. It was the introduction of this type of program, in league with cost-effective PCs and laser printers, that made it possible for the average computer owner to participate in the DTP field.

A page-composition program produces a display similar to that of a WYSIWYG word-processing program with numerous enhancements. Software tools are used to accurately place text and graphics, and there are alignment aids, such as a snap-to guide, that can line up these elements on a page.

A page-composition program supports a host of other features, including text- and graphics-manipulation modes, and in many cases, an extensive library of templates. The discussion that follows provides a general overview of these operations.[13]

Text Manipulation. When you create a new document, the program will display an outline of a blank page that represents the first page of, in our example, a newsletter. Next, the computer can open up one or more columns on a page if this function is selected.

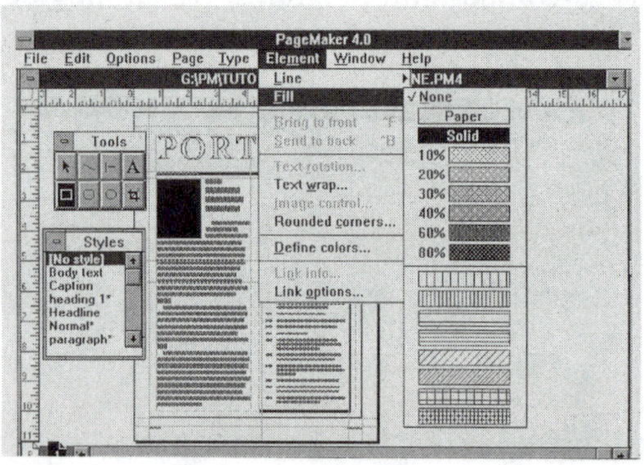

The text and images that compose the newsletter are placed in the columns and in resizable, movable blocks or frames.

Our newsletter is formatted like a newspaper. A page consists of columns of text and assorted illustrations, and the top of the first page displays the newsletter's masthead. To create this last space, a mouse is used to open up a long and narrow frame across the top of the page. The size and typeface of the masthead's characters can then be selected, as can other elements of the text's physical characteristics. These include printing the text with a shadow effect, in a different typeface than the rest of the document, or in a larger point size.

A DTP program is also equipped with a word-processing module. In most cases, though, you will continue to use your word-processing program of choice to complete the initial draft, since it may be more sophisticated and faster than the one provided by the DTP software.

DTP programs can also import text files from many of the most popular word-processing packages. The file is retrieved by the DTP system and, in one operation, can be placed in columns that have been reserved or designated for the text.

If the file is large and one column fills up, the text can be routed to another column. The routing can be automatically controlled by the computer, or you can manually designate the next column the text should fill.

A program can also compensate for the addition and deletion of text. It is a dynamic system that can automatically adjust for editing changes. As words are added or deleted, the text flows or snakes from column to column until the proper space adjustments are made.

This control over the text also extends to the physical spacing between individual characters and sentences. In *kerning* and *leading*, respectively, the space between specific pairs of letters and between individual lines of text can be altered. The capability to vary the blank or white space can create a good balance between the type and white space on a page. It may also make it easier to read the document and to fit additional text in a physically smaller area.

Figure 7-6
Some of the many options and tools supported by DTP software. (© Aldus Corporation 1987–1990. Used with the express permission of Aldus Corporation. Aldus® and PageMaker® are registered trademarks of Aldus Corporation. All rights reserved.; PageMaker)

Beyond these features, page-composition programs can now import a greater number of file types. These include data from different graphics, word-processing, and spreadsheet programs. It may also be possible to produce special effects, such as rotating a line of text for vertical placement on a page.

The trend appears to be the creation of more self-contained programs. As various modules are added and refined, including the word-processing function, a page-composition program may be able to handle more tasks without using other software.[14]

Graphics Manipulation. The integration of a graphic on a page of text is one of the more powerful features of DTP software. Most DTP programs can create very simple graphics, but like the word-processing function, the capability may be so limited that most people will want to use one of the dedicated graphics programs described in Chapter 3 for this task.

Once a graphic is created—be it a line drawing or an artist's rendition of a mountain (which will illustrate an article in our newsletter)—it is imported by the DTP package. At this point, the graphic's size can be scaled or altered. The graphic can also be *cropped* so only a portion of the entire image appears.

The graphic can then be moved to other positions as space permits. The DTP program may also support a wraparound feature where the text in the document flows or wraps around the image.

Templates. A DTP program may be packaged with a series of templates. A template specifies the design of books, magazines, newsletters, and other documents, and is a valuable publishing tool.[15] Rather than spending hours of time designing a format for a publication, you can use the template to complete this task and delineate the document's physical appearance so it conforms to an accepted standard. Custom templates can also be created, stored, and recalled when necessary.

Besides helping novices, the template can be useful to people who are in a hurry or who cannot design an attractive, yet practical, format. While a DTP program does simplify this task, it may not be an easy process.

A premade template can help eliminate this problem. The document will be visually more appealing and should be easier to read. A template can also bring a sense of order and uniformity to reports. A large company, for instance, can house a number of divisions that produce their own written reports. If a format is not adopted, a report's structure could vary widely from division to division. This could potentially impede the flow of information in the company, especially if a document does not present the material in a clear and logical fashion.[16]

Finally, it's appropriate at this point to discuss several basic guidelines that can further enhance a DTP operation.

1. Institute a training program if a DTP configuration is adopted. A program may be very complicated and have a steep learning curve. An individual may have to spend hours working with the software before he or she is conversant with its major functions.

2. A DTP system will not turn everyone into an artist. It is simply a tool that can be used to present ideas and information more effectively. We still need artists to draw graphics and individuals to design appropriate documents.

Paraphrasing Clint Eastwood in one of his Dirty Harry roles, you've got to know your own limitations. In this case, it also includes your strengths. Do what you do best, but if you need a professional, hire one. It will save time and money and will most likely result in a superior final product.

3. Plan and effectively use the white space on a page. White space can provide visual relief for the reader and serves as a design tool by highlighting or focusing attention on specific page elements. This concept extends beyond the DTP industry and, in fact, is an established technical-writing axiom.

4. Pay attention to the basics: grammar, typos, and spelling mistakes. Proofread the document after you're finished, and don't try to catch every kind of mistake in only one reading.

5. When designing a document, keep it

simple, if appropriate. While a DTP program may support an enormous range of typestyles and special printing effects, don't use them all on the same page. Your new document may be very difficult to read, and in fact, the information may be lost in a maze of fonts and justified double-underlined text.

6. Read and practice: there's a wide range of DTP books and magazines that cover everything from aesthetics to scanning. A magazine may run a series of articles that explores practical skill areas, while others may be compilations of tips from professional DTP users.

7. If you're learning how to use a DTP configuration, it's also important to practice your craft. Experiment with what you've learned and develop your own style.

8. Also, use your imagination. If you have a tight budget and deadline, don't have access to color equipment, and have to complete a project with a color image, you may be able to use a variation of the cut-and-paste method. Create the page with a DTP program and then use the original color image, in the appropriate space, with one of the new color copying machines for the output. Depending upon the job, the result may be satisfactory.

APPLICATIONS AND IMPLICATIONS

Personal Publishing

You can use a DTP system in a number of applications, one of which is as a personal printing tool to publish your own book. Prior to this time, if you wanted to pursue this goal, you generally had two options. You either signed a contract with an established publishing house, which would print and distribute the book, or you used a vanity press. In the latter situation, you paid a publisher for printing the book, and the company usually handled various distribution duties.

The introduction of cost-effective DTP outfits has provided a third alternative. You can now act as your own publisher and distributor.

In this operation, the author is in control of the publishing process. Last-minute changes

and updates can be accommodated, and a standard laser printer may suffice as the electronic printing press. Or a higher-dpi commercial unit can be used to produce the final copy, which is subsequently duplicated for distribution.

In another variation of this theme, you can initiate what Don Lancaster has called a *book-on-demand* publishing operation. You do all the work yourself, including the printing and binding. But instead of producing an initial run of 500 or more copies, you print a book only when someone orders it. In this way, the up-front costs for materials or a printing service aren't as great, and each copy could literally be an updated version of the original book.[17]

While both methods would allow you to be your own publisher, there are some constraints. They vary from individual to individual and include your initial budget, experience, and DTP system's level of sophistication. These factors will have an impact on your final product.

You must also face the dual problem of promoting and distributing your work. While the structure for this market is evolving, it may still have to mature to provide a more established support mechanism. You also don't have access to the editorial and technical expertise afforded by a traditional publishing house.

Other Applications

Publishing companies have also used DTP technology. Since DTP systems are cost-effective, smaller institutions could potentially publish more books and take chances with manuscripts that are geared toward narrow and potentially limited audiences. Larger companies could use the technology with the same goals in mind.

Beyond these applications, DTP systems have been adopted by organizations for a wide range of jobs. These include the publication of technical manuals, year-end reports, ads, information flyers, posters, and newsletters.

Since desktop-publishing systems can save time and money, they have also been used by more traditional media organizations. *The*

New Yorker, for one, slowly integrated desktop-publishing technology in its operation. The move was initiated to speed up certain operations and, as stated, to save money.[18] In the newspaper industry, PCs equipped with the appropriate software have even been used for various photographic editing and preparation tasks.[19]

Individual users who want to create near-typeset-quality documents, regardless of the product, have likewise benefited from DTP technology since printing and copy shops have set up in-house DTP stations. In a typical scenario, a shop may be equipped with a PC connected to a laser printer. A customer can then have a résumé or other document produced with this system, typically at a lower cost than the traditional typesetting route.

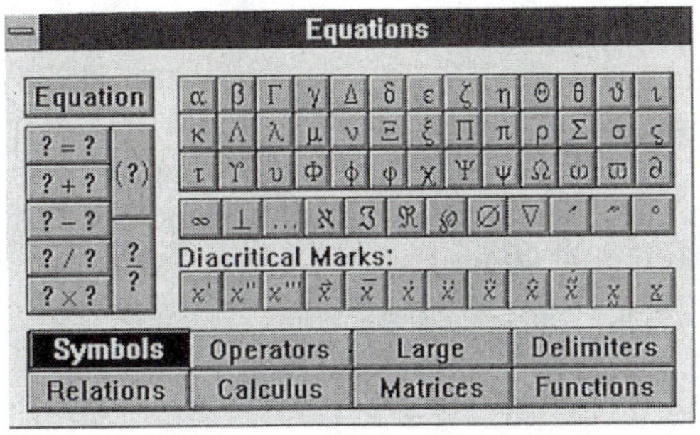

Figure 7–7
One such capability that enhances the production process is a series of tools that can help simplify what may normally be a complex task: in this case, the capability to generate and use a range of equations via an equations tool. (Reprinted with permission of and copyright © 1986–1992 Frame Technology Corp. All rights reserved; FrameMaker)

clip-art collections are sold precisely for this purpose, to serve as illustrations.

If either of these solutions does not solve the problem, it may be possible to hire an artist to create the graphic. Or for a straightforward image such as a simple chart, a graphics program would enable an individual with even marginal artistic abilities to produce an acceptable image.

Legal and Ethical Implications

While computer technology has provided us with an incredible array of personal publishing tools, the same technology has also created a legal and ethical problem with respect to copyright law and intellectual property rights.

Scanners, for example, have made it possible to copy a graphic produced by another individual and to use it in a publication without the artist's permission. The same graphic can also be scanned, altered with either a graphics or image-editing program, and then used in a publication. In both cases, the artist's rights and the copyright law are violated. But, as discussed in Chapter 3, the widespread integration of PC-based tools in all levels of society has made it almost impossible to protect the artist's rights in this type of situation.

There are, however, legal ways to use preexisting graphics in a publication. In one case, it may be possible to obtain the proper clearance from either the artist or the organization that owns the rights to the graphic. The clearance may be granted for free, if it is for an educational or nonprofit endeavor, or as will most likely be the case, it may be used for a specified fee. In addition to this option,

The Democratization and Free Flow of Information

In another application area, DTP technology has contributed to what can be called the *democratization of information*. An individual with the requisite skills and money can buy a DTP outfit and can publish a newspaper, magazine, or a book. This capability implies that information cannot be controlled by only a select group of people or by the government.

A government could, for example, censor or even shut down newspapers, television stations, and other established media organizations. But it may be impossible to completely shut off the flow of information in a society when there are literally thousands of smaller printing presses in the form of DTP systems and the means to reproduce the publications via copy machines.

This concept has been put to the test on different occasions, including the aborted August 1991 coup in the Soviet Union when Mikhail Gorbachev was deposed. Desktop-publishing systems and fax machines played a role in this dramatic event.

Russian President Boris Yeltsin and his support group used these communications tools, and others, to keep the Russian citizens and the world community apprised of events.

The free flow of information, during this time, contributed to the coup's eventual collapse.[20]

The democratization of information also implies, in the context of our discussion, that more individuals and organizations have access to a powerful publishing tool. The applications vary widely, from the business world to personal publishing, and the end result is the capability to produce your own work. If applicable, it can even be placed in the open marketplace.

No one may actually view or, in this case, read your work. But even if this holds true, you can still use a DTP system to express your ideas and to present them in the form of a letter, newsletter, book, or even a political broadside.

Finally, the idea of using a new communications technology to support the free flow of information is not limited to this industry. As described in various sections of the book, different technologies and their tools can and have served a similar purpose. These include satellites, computer networks, and desktop video systems.

CONCLUSION

As of this writing, the DTP industry is still growing and maturing. Besides the developments outlined in this chapter, the DTP industry will also benefit from the convergence of different technologies and applications.

Entire font and clip-art collections can, for instance, be stored on CD-ROMs. There is also an overlap in the area of graphics software. As briefly described in Chapter 3, a CAD program can be used to design a building. In a desktop-publishing or video project, the same program can create an illustration.

Advancements in one field can also have an impact on another field in the new technology universe. Desktop publishing is no exception.

Newer, powerful PCs speed up various operations, and for DTP, a project can be completed in a shorter time frame. These same machines can also support sophisticated software that were once the exclusive

domain of larger and more expensive computer systems. In one application, DTP users can correct and enhance images for a publication. Similar techniques have been and are used by the medical and scientific communities for their particular applications.

In two other examples, progress in the overall laser market will have an impact on DTP systems (for example, laser printer development), while the proliferation of networks could promote applicable DTP-based operations. In this situation, an electronic-publishing environment would make it possible for more than one person to retrieve, review, and edit the information that composes a document as well as the document itself.

The network, though, should be capable of handling a large volume of information. It should also be equipped with a fast and large data-storage system to accommodate multipage documents, which are typically integrated with graphics.

The concept of networking also cuts across national and international boundaries. As new high-speed digital lines enhance the communication process, offices around the world can be tied in a global communications network. For DTP operations, you can gain access to the data stored on other networks, information can be rapidly exchanged, and ultimately, important resources can be shared.[21]

The DTP field has also helped promote the growth of a personal communications tool. With a DTP system, an information consumer can now become an information producer, as described in the applications section of this chapter. Information can also be tailored for a narrow rather than a mass audience, as may be the case with newsletters, pamphlets, and even book-on-demand publishing projects.

These ideas have been extended to the electronic media. The next chapter explores desktop video and multimedia systems. Much like a DTP application, an individual or organization can produce a publication. But in this case, the final product is in an electronic form that can be viewed on a monitor and recorded on videotape.

REFERENCES/NOTES

1. A hard drive, in this configuration, can speed up a printing operation.
2. Tom Thompson, "Color at a Reasonable Cost," *Byte* 17 (January 1992): 320.
3. These printers also produce a continuous-tone output, in contrast to halftone-based systems. Halftoning is described in a later section of this chapter; continuous-tone images look more like conventional photographs. Please see Tom Thompson, "The Phaser II SD Prints Dazzling Dyes," *Byte* 17 (December 1992): 217, for specific information.
4. John Gantz, "DTP Is Inching Toward Color, but Don't Hold Your Breath," *InfoWorld* 13 (June 10, 1991): 51. Also see Janet Anderson, Philip A Borgnes, Carol Brown, et al., *Aldus PageMaker Reference Manual* (Seattle: Aldus Corporation, 1991), 76–83, for an excellent overview of color printing via a DTP system.
5. If you decide to go the commercial route, you can use process- or spot-color printing. See Eda Warren, "See Spot Color," *Aldus Magazine* 3 (January/February 1992): 45, for more information.
6. Daniel J. Makuta and William F. Lawrence, *The Complete Desktop Publisher* (Greensboro, N.C.: Compute! Publications, Inc., 1986), 86–87.
7. This is especially true when using one of the software PostScript interpreters.
8. Image-In, Inc., *Image-In* (Minneapolis: 1991), 178.
9. An in-depth discussion of gray-scale representations, scanners, and laser printers is featured in a special issue of *Personal Publishing*. Two articles in the May 1988 issue, "Impressions of Gray" by Steve Roth, pp. 24–34, and "Gray Scale Tradeoffs" by Patrick Wood, pp. 36–37, 41, describe the entire digital halftoning process in addition to tips on how to produce superior print copies of original photographs.
10. This type of technique has been employed for a number of years in various applications. They range from the digitizing systems described in the next chapter, where an image from a videocamera can be produced in a color format, to the color pictures produced by NASA's outer-space probes.
11. Barbara Bourassa, "Mac Systems Speed Photo Transmissions," *PC Week* 9 (February 17, 1992): 25.
12. In addition to word-processing and page-composition software, another program category has offered a number of page-formatting or design options. A series of codes is used, analogous to the codes embedded in a document produced by a word-processing program, to implement a specific page design. Programs of this type have been written for professional computer typesetting systems, while PC software also exists.

    The text produced by a format program may not appear on a monitor in a WYSIWYG format, as described in the next section of the chapter, and the various control codes are not implemented until the printing actually takes place. But a program may offer a print preview mode, in which the pages in the document are displayed as they will be printed.

    The advantages offered by this type of program include a superb formatting capability. A program could also be used with a range of printing equipment, and some programs routinely handle very difficult formatting tasks.
13. Please note that the terms used can also vary from program to program even though the basic concepts hold true.
14. As previously described, word-processing programs have similarly been enhanced. One price, though, for the increased sophistication of both types of programs may be software that is basically bloated, too powerful for simple tasks. A program may also place a higher processing and data-storage demand on the host PC.
15. Virginia Rose, *Templates Guide* (Seattle: Aldus Corporation, 1990), 3.
16. When working with text, it's also possible to use a style sheet. In essence, a style sheet defines the attributes of a document's different elements, such as a headline and body text. A headline may be centered and set in a specific typeface and size. To use this style, highlight the appropriate word(s) during editing, select the headline style, and the text will be automatically reformatted.
17. Don Lancaster, "Ask the Guru," *Computer Shopper* 9 (September 1989): 242.
18. James A. Martin, "There at The New Yorker," *Publish* 6 (November 1991): 53.
19. Jane Hundertmark, "Picture Success," *Publish* 7 (July 1992): 52.
20. Please see Richard Raucci, "Overthrowing the Russians and Orwell," *Publish* 6 (November 1991): 21, and Howard Rheingold, "The Death of Disinfotainment," *Publish* 6 (De-

cember 1991): 40–42, for additional information about this event and similar incidents, such as the 1989 Tiananmen Square uprising. Rheingold's article also touches on other implications raised by the new communications technologies and the free flow of information.

21. Lon Poole, "Digital Data on Demand," *MacWorld,* February 1992, 227.

ADDITIONAL READINGS

Publish 7 (August 1992). This issue of *Publish* has three articles dealing with color printing. They range from how to generate color proofs of your work to calibrating your monitor for color work.

Fraser, Bruce. "Color Code." 53–56.

Yi, Paul. "Screen Tests." 59–62.

Hannaford, Steve. "All the Proof You Need." 64–68.

Aldus Magazine 3 (January/February 1992). The following three articles are concerned, respectively, with selecting a paper type for DTP, the history of paper, and a history of offset lithography.

Beach, Mark. "Paper in the Short Run." 33–36.

Stratton, Dirk J. "Down the Paper Trail." 80.

Vick, Nichole J. "Oil and Water." 19–22.

Alford, Roger C. "How Scanners Work." *Byte* 17 (June 1992): 347–50. An excellent overview of scanners and their operation.

Berry, Sarah M. *Aldus FreeHand and Commercial Printing.* Seattle: Aldus Corporation, 1991. While this publication is a manual for Aldus FreeHand, it also serves as an introduction to working with commercial printing as well as color.

Bishop, Philip. "Crimes of the Art." *Personal Publishing.* May 1990, 19–25.

Parker, Roger C. "Desktop Publishing Common Sense." *PC/Computing.* March 1989, 151–56. And, "Desktop Quality Circa 1992." *Business Publishing* 8 (January 1992): 23–29. "Publish Special Section; 101 Hot Tips." *Publish* 7 (July 1992): 63–88. Desktop-publishing tips and guidelines: All four articles provide a comprehensive overview of DTP aesthetics and how to effectively communicate your message.

Fraser, Bruce. "Moving Up to Level 2." *Publish* 6 (November 1991): 78. And, Gass, Linda, John Deubert, et al. *PostScript Language Tutorial and Cookbook.* Reading, Mass.: Addison-Wesley Publishing Company, Inc., 1985.

Lancaster, Don. "Ask the Guru." *Computer Shopper.* And, Pelli, Denis G. "Programming in PostScript." *Byte* 12 (May 1987): 185–202. The references provide information about the PostScript language, ranging from a discussion of language enhancements to programming techniques. The Lancaster reference has been a column of articles covering, among other topics, PostScript printers and the PostScript language. Don Lancaster is a PostScript authority and offered in his column a number of tips and fonts readers could use in their own work.

Pennycock, Bruce. "Towards Advanced Optical Music Recognition." *Advanced Imaging* 5 (April 1990): 54–57. The use of PCs and scanners for automatically recognizing musical scores.

Rosch, Winn L. "Power Printing." *Computer Shopper* 12 (July 1992): 584–94, 598, 600A. A comprehensive overview of printers, how they work, and the questions you might ask when selecting an appropriate unit. Also contains sidebars by Gregg Keizer on fonts and color printing.

Salpeter, Judy. "Are You Obeying the Copyright Law." *Technology & Learning.* May/June 1992, 14–18. An overview of copyright issues that is especially relevant for schools. Also lists names of associations concerned with copyright issues.

Seybold, John, and Fritz Dressler. *Publishing from the Desktop.* New York: Bantam Books, 1987. An excellent survey and analysis of the development of the DTP field, among other issues.

Shushan, Ronnie, and Don Wright. *Desktop Publishing Design; Aldus PageMaker Edition.* Redmond, Wash.: Microsoft Press, 1989. Although geared for the PageMaker DTP program, the book provides a comprehensive view of DTP design and aesthetic issues, and includes hands-on projects.

GLOSSARY

Color printers: Different color printers support the general consumer/business markets. They range from ink-jet to thermal-wax units. More expensive and sophisticated printer types exist as well.

Desktop publishing (DTP): A term that describes both the field and the process whereby near-typeset-quality documents can be produced with a PC in combination with a laser printer and, typically, special software. DTP also implies that a user has access to a variety of layout and printing options as well as the capability to finely tune a document's appearance.

Font: A font is a typeface in a specific size.

Laser printer: One of the primary components of a DTP system. A typical printer produces near-typeset documents and can support a 300-dpi output (newer models support a higher dpi figure).

Optical character recognition (OCR): Either a stand-alone unit or a software option for a scanner that makes it possible to directly scan the alphanumeric information from a printed page to a computer.

Personal publishing: Desktop-publishing systems make it possible for individuals to produce and potentially market their own work (for example, books).

PostScript: A device-independent page-description language (PDL) that is a popular and flexible DTP standard.

Scanner: An optical/mechanical device that can be interfaced with a computer. Either desktop or hand-held scanners are used to scan and subsequently input graphics or text to a computer.

Typeface: A specific and unique print style.

8 Personal Media: Multimedia and Desktop Video

As defined in Chapter 4, the term *multimedia* can describe the integration of graphics, audio, and other media in a single presentation. This chapter explores the dedicated multimedia program, a software package designed to create and control such a production. Other relevant topics include complementary hardware components, applications, and aesthetic considerations.

The chapter also covers the desktop video field. Advancements in video technology have made it possible for many individuals to create their own video productions, and an integral element of this process is the computer. PCs are used for applications ranging from editing to creating graphics.

The desktop video and multimedia fields are also complementary in nature. Desktop video tools can be used to create various elements of a multimedia presentation. Desktop video, in turn, can be categorized as a component of the broader multimedia market.[1]

MULTIMEDIA

The concept of a multimedia presentation is not new. The videodisc, for example, in conjunction with software that can tap this audio-visual resource, has served as a multimedia platform for a number of years. In a typical application, information from the videodisc can be viewed in a special window on a monitor's display.

While the videodisc still plays a major role in this market, other products, discussed in the next two sections, have also become popular and have contributed to the field's steady growth during the 1990s. Some of these factors include

- the introduction of powerful PCs;
- the release and broader adoption of new software packages;
- the low cost of computer memory;
- developments in the optical-disc field;
- the growth of the desktop video field; and
- the ability to use digital audio and video in a presentation.

These developments, among others, have helped shape the contemporary multimedia market.

There are other multimedia resources in addition to the ones discussed in this chapter. These include a CD-I title that may serve as a stand-alone multimedia product, as well as other members of the optical-disc family.

Certain software packages also fall into this category. A hypermedia program's primary goal is the creation of a hypermedia document. But as described in Chapter 4, it may also support different media. Other program

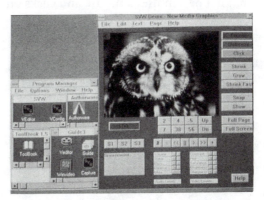

Figure 8–1
With the addition of a card, for example, a PC could play full-motion video in a resizable window (on a monitor). (Courtesy of New Media Graphics; Super VideoWindows)

types may similarly be able to create their own multimedia presentations.

Operation

Software. In the context of our discussion, the software used to create a multimedia presentation is an *authoring* program. Many authoring systems are user-friendly in that you don't have to be a programmer to create a finished product. One package, Amiga Vision for the Amiga PC, has used a visual metaphor for this task.

A series of icons, which represent different functions, are linked together to create the presentation. The icons are used as audio-visual and program-control building blocks. Thus, one icon may be used to play back an animation at a specific point in the presentation. Another icon may be used to create a loop, which when reached causes a series of events to be repeated. In this case, the animation may be replayed.

When applicable, an icon is defined by the programmer. In one example, by clicking on an animation icon, a *requester* or *dialog box* may appear. The box prompts you to select the specific animation you want played.

A program may also support a wide range of transitions between the various elements that compose the multimedia production. These could include a fade-to-black, a wipe, or a dissolve to the next image. Transitions, if used properly, can help the flow of a production while enhancing its aesthetic value.

Besides its ease of use, an authoring system can create a highly interactive presentation. If a person uses a mouse and clicks on a specific area of a displayed graphic, an event may be triggered through a preset link. A digitized voice that describes the graphic could be heard or a clip from a videodisc could be played.

In addition to icon-driven authoring software, a multimedia package may support other types of programming interfaces, such as a text-based scripting language. Asymetrix Corporation's ToolBook, for one, released when Windows 3.0 was launched into the software market, has employed the Open-Script programming language to create complex multimedia projects, among other applications. These can include hypermedia and interactive educational applications. A program of this nature can also construct a user-friendly front end, an interface, to a database.

While languages such as OpenScript are fairly easy to learn, it's possible to create an application without programming per se. ToolBook is also equipped with a library of prescripted multimedia objects, such as control panels, to help simplify and speed up the production process.

Beyond these characteristics, authoring systems are typically more flexible and powerful than conventional programming tools for specific data-handling tasks. The seamless integration of various audiovisual elements in a multimedia presentation, and the control of videodisc players and other hardware components, are two such examples.

The key word in the last sentence is *seamless*. Unlike a conventional language that may require add-on software modules or special programming hooks for the operations, the capability is built in and is fully integrated in multimedia authoring packages. But a price for this ease of use may be a reduction in operating speed. An authoring system may not complete specific functions as quickly as a program written with a conventional language.

While this type of software may be easier to use than standard programming tools, you must still abide by one programming convention. To sustain an effective multimedia environment, the presentation must flow logically from event to event.

Initially, a set of criteria that will drive the

Figure 8–2
The list of supported prescripted objects. (Software courtesy of the Asymetrix Corp.; Multimedia ToolBook)

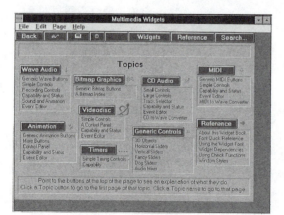

program's design should be drafted. The criteria should include the program's purpose and the reason for its development. Next, determine the best way to satisfy these criteria with the software's programming tools.

Hardware. Multimedia presentations are supported by Apple, IBM, and Amiga PCs, among other computer systems. In addition to the built-in hardware capabilities, a PC platform may require additional hardware peripherals to take full advantage of a multimedia environment.

One such product is a CD-ROM drive. A multimedia presentation may be composed of high-resolution graphics, animations, video clips, and high-quality audio. The volume of information may mandate the use of a storage system capable of handling this much data. A CD-ROM designed for this application serves as an ideal storage mechanism.

Besides a CD-ROM, a board that provides an enhanced audio capability may be required. This is especially true for many stock IBM PCs that have, as of this writing, only rudimentary audio capabilities.

To tap this power, which may also be built in depending upon the PC platform and function, audio software can be used. You can digitize your voice, view it as a waveform, and subsequently edit and alter it. You can select echoes and other special effects, and you may be able to mix your voice with music. In this case, the software provides you with a computer-based version of an audio console.

Unlike a traditional audio setup with which you cut and splice tape, the work is performed on and with the computer. You use a mouse, in conjunction with the monitor's display, to select the different functions and the audio clip you want to alter. If you save the original file, it's possible to experiment with and save different variations of the same piece.

Sound effects, such as a bell ringing or a beep, are also common, and can serve as useful audio cues. In a multimedia presentation, a sound effect can provide a user with feedback confirming the selection of an on-screen button that triggers an animation or other event.

Like clip art, it's also possible to buy music clips. Basically, you don't have to be a musician to take advantage of audio cuts in your production. This is an important consideration, since audio, when properly integrated, can help make a presentation more effective and interesting. Multimedia presentations can also take advantage of the MIDI protocol for the control and playback of other audio sources. MIDI systems are described in detail in a later section of this chapter.

Beyond these devices, the primary hardware consideration is the PC itself. The creation of a multimedia presentation is typically a hardware-intensive operation, regardless of the PC platform.

In general, a PC's performance is an important concern. While it's possible to create a presentation with most machines, you should use the most powerful PC you can afford. If not, you could waste valuable time waiting for the computer to complete even a single operation. If you are working under a tight deadline, this may be unacceptable.

The PC should also be equipped with a fast, high-capacity hard drive. As previously indicated, the types of information that may be used in a presentation are storage intensive. Even two or three "average" image files could require well over a megabyte of storage, and digitized video and audio are even more demanding. Thus, a large hard drive, which is also fast enough for quick storage and playback operations, is a necessity. Erasable optical-disc systems should also prove valuable, especially as they become less expensive and their performance continues to improve.

Another hardware requirement is a suffi-

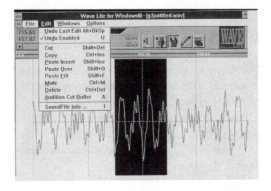

Figure 8–3
PCs, when combined with software, can create an audio-editing system. You can now view the pertinent audio cut and edit/manipulate a specific section (which appears highlighted in this shot). (Software courtesy of Turtle Beach Systems, Inc.; Wave for Windows)

cient amount of RAM: the more RAM the better. Besides meeting minimum memory requirements to run a program, additional RAM can speed up certain operations.

The video display should also be considered. The new generation of video cards can support lifelike, that is, photorealistic, images. They may also offer an accelerated operation either by themselves or through a special connection. The IBM-PC world, for example, only recently (in the early 1990s) benefited from local bus video. Essentially, a special slot will speed up a PC's video performance or throughput.[2]

Finally, Apple and Amiga computers were well ahead of the IBM-PC world with respect to hardware-compatibility issues. Stock systems were also able to support a multimedia operation more effectively than an off-the-shelf IBM PC.

But the scale started to balance with the introduction of Microsoft's Windows 3.0 and the optional Multimedia Extensions. The Windows environment, despite a high-performance hardware demand, offered developers and users a standard operating interface. The Multimedia Extensions, in a complementary fashion, provided the necessary interface for hardware compatibility.

The next Windows release, Windows 3.1, was multimedia ready. The program also included a Media Player, to "play multimedia files" and "to control hardware devices," such as a videodisc player.[3] These developments led to the introduction of multimedia hardware-upgrade kits as well as IBM PCs equipped with suitable CD-ROM drives and other multimedia components. Windows was also able to accommodate digital video, a topic described in a later section.

DESKTOP VIDEO

The term *desktop video* means basically what it sounds like. You can now set up a video production system on a desktop.

Desktop video is also concerned with computer technology. PCs can control video equipment and play a major role in the video production process.

Desktop video products range from consumer-grade to professional equipment. Over the last few years, there has been a trickle-down effect from the professional to the business and consumer markets. Equipment became more sophisticated while market competition and other factors drove prices down. Eventually, some of the lines separating the professional and nonprofessional markets became fuzzy.

Similarly, various desktop systems described in this chapter have found a home in both markets, and different levels of PC-based systems have emerged. In one example, desktop video now supports editing operations, which can be used by an individual, a production house, or a network. But the sophistication, speed, and durability of the systems vary, even though the basic concept, the capability to create a production, remains.

The steady growth of the desktop video market has also made it possible for the average person or organization to create a video production. It can range from a video home movie to a series of animations recorded on videotape to a component of a multimedia presentation.

This development is a major step in the transformation of mass media to personal media. In a complementary fashion, it also empowers people. For the first time in the history of modern communications, individuals and smaller organizations can now possess the information tools that were once the exclusive domain of established media groups. As discussed in the section on the democratization of information in Chapter 7, developments of this nature have important implications for the free flow of information in a society.

Operation

This section covers the basic equipment and software used in desktop video applications. While it's important to examine the capabilities of individual elements, it's also important to view them in the context of the overall production process. They work as part of a system.

A component may also have multiple applications: a PC may be used for creating graphics as well as editing.

Frame Grabbers, Slow-Scan Digitizers, and Motion Video.

A digitizing system usually consists of a camera and an interface device, which may be an internal PC card. Much like a scanner, the system converts images into a computer-compatible format. Unlike typical scanners, though, a camera can work with three-dimensional objects.

A frame grabber can capture and digitize an image in real time, in 1/30 of a second, the standard video frame rate. In a typical application, you feed a videocamera's signal to the frame grabber. A specific frame, essentially a still image from the video, is captured by a click of the mouse button or a keystroke.

Slow-scan digitizers, on the other hand, may take several or more seconds to complete their capture process. This generally limits their use to stationary objects. But these units are typically less expensive.

Once the image is captured, regardless of the system, it can be saved as a file in an appropriate graphics format. At this point, the image can be manipulated by a graphics program.

Finally, other cards allow you to capture motion video, moving images. This capability plays an important role in the PC-based digital systems described in a later section of this chapter.

Encoders and Genlocks.

An encoder can convert a PC's output, the signals used for the display, into an NTSC format. The encoder makes it possible to record an animation or other graphics display on a VCR.

A genlock carries this capability one step further. A genlock can sync and mix a computer's output with external video. In one application, you can use a genlock to overlay computer-generated text, the title for a production, over the video.

Encoders and genlocks are important devices since they allow you to record your work on an easily distributable medium, videotape. But, like desktop publishing, there are some basic conventions you should follow when creating a project for a video environment. A few of the important ones include the following.

1. Check all your work on a television or monitor after it's converted or, even better, while you're working. There's a drop in quality during the conversion process. Colors may not be as rich and there's a loss in resolution since the fine details visible on a computer monitor may not be visible on a TV screen.

2. Don't place important visual elements, such as titles, right at the edge of the screen. They might be cut off when displayed on a television.

3. Make sure the text isn't so small that it is hard to read. You should also avoid a fancy script typeface. While it might be fine on the computer's display, it might be illegible on a standard television.

4. Don't set color values at their highest settings. The DeluxePaint program for the Amiga PC, for example, lets you control the color palette, the selection of colors, with an RGB slider control. Avoid topping off a slider and/or the sliders at the highest setting since the resulting colors may not be cleanly reproduced when transferred to video.

5. When recording, don't use old tape. Videotape that has been reused a number of times may not give you a clean recording.

6. Be paranoid. Always save your work and back it up. This is standard practice for any PC activity. For graphics work, where you can easily spend hours to create an animation or even a single image, make this procedure a habit.

Edit Controllers.

PC-based edit controllers are used to control editing VCRs. Editing is the process in which an extended series of shots that compose a story is electronically assembled by selecting, organizing, and joining the individual shots. The final edited tape consists of one shot followed by the other.

The systems used for desktop video applications can range from simple *cuts-only* systems, where shots are simply linked, to *A/B roll* configurations, where you can make more sophisticated transitions. For example, instead of just cutting between shots, where the transition takes place instantaneously, you can use a dissolve, which gradually replaces one image with the next. This type of setup can be complex and requires additional equipment, including VCRs and a device to create the transitions.

An editing system has also become a realistic production option for desktop video users. If you already own the computer and VCRs, a basic system could cost only a few hundred dollars. While this setup may support only simple operations, it nevertheless gives a broad user group the power to edit their own productions. Different videotape formats may also be accommodated (for example, S-VHS).

Musical Instrument Digital Interface. In addition to the audio boards described in the multimedia section, PCs can take advantage of the Musical Instrument Digital Interface (MIDI) standard. In fact, for IBM PCs, an audio or sound board that expands the computer's audio capabilities may also be equipped with a MIDI interface.

The MIDI standard made it possible for electronic musical instruments developed by different manufacturers to communicate with each other and, ultimately, with computers. This interface is somewhat analogous to the MAP standard discussed in Chapter 3, since they both link a wide range of equipment in a single communications-and-control network. But the MIDI standard was developed for the musical rather than the industrial world.

The MIDI interface provides an individual with a powerful composing tool. In a typical application, a computer is directly linked with a synthesizer, an electronic musical instrument. This connection is made through the PC and synthesizer's MIDI ports, and a program can subsequently turn the computer into a *sequencer*. A sequencer essentially is a multitrack machine that can record and play back multitrack compositions.

You can manipulate the notes played on the synthesizer with the PC in various ways, since these notes, the events, are viewed by the computer as another form of data. The actual sounds are not recorded, but rather, information detailing the performance. The

"speed at which the key was pressed" is one such piece of information.[4]

With a MIDI configuration, you can create and edit musical compositions. A section of a track can be copied or deleted, the key can be transposed, and the tempo can be altered, among other options. The final piece can then be played back under the control of the PC.

The computer–MIDI marriage offers other advantages. A synthesizer can create a range of sounds that can vary from a harpsichord to a pulsating tone, a special effect that may be suitable for a science-fiction movie. The parameters that constitute a sound can also be saved on a disk, and it may be possible to store, manipulate, and recall entire sound libraries. Collections of premade sounds are also available.

The MIDI revolution was brought about by the adoption of the standard in the early 1980s. Electronic-equipment manufacturers agreed to follow this common standard to enable musicians to link multiple synthesizers, helping to fuel the growth of the electronic music industry. A wide array of previously incompatible and expensive instruments could now be interfaced, and musicians were handed a set of creative tools that enabled them to control the manipulation of sounds and the production of musical compositions. The musicians, in a sense, could use a PC and the appropriate software to "paint" a musical piece just as graphic artists could paint multicolored pictures.

The industry's growth was enhanced by the expanding PC market and the introduction of MIDI interfaces and software. Another contributing factor was the drop in price for the instruments and PCs.

Besides the development of this element of the electronic music industry, manufacturers have introduced inexpensive sampling keyboards, fairly sophisticated products that support lower-end applications. This instrument is suitable for an individual who wants to experiment with electronic music but has either no inclination or no money to purchase a full MIDI system.

The typical keyboard can store and replay a set number of notes, has an internal library of sounds or instrument emulations (for example, piano), and can digitize external sounds. This includes a human voice that can subsequently be manipulated and replayed as a part of a musical composition. While the instrument cannot match the capabilities of one of the more expensive units, it does introduce the user to the concepts that govern the electronic music field.

Compression: Overview. The term *compression*, for our purposes, refers to data compression. Digitized video, audio, and certain image files have enormous storage appetites. By using various software and/or hardware schemes, data can be compressed and stored more efficiently.

Compression can be *lossy* or *lossless*.[5] *Lossy* means some of the data are lost through the compression scheme. For certain applications, such as saving still images for desktop publishing, this may not be a problem, since the loss can be negligible. *Lossless*, on the other hand, implies there isn't a loss of data. Lossless techniques are used in application areas such as the medical field, in which a data loss may be unacceptable.[6]

Popular compression schemes include the *Joint Photographic Experts Group* (JPEG), the *Moving Picture Experts Group* (MPEG), and by extension, MPEG-2. The JPEG and MPEG standards were originally and nominally geared for still and moving images, respectively.

Another standard, *Digital Video Interactive* (DVI), was brought to the desktop by the Intel Corporation in league with IBM. Originally developed by RCA, the DVI system is centered on a highly efficient data-compression technique. An impressive achievement when first announced in the 1980s, it can support a full-motion, full-screen display. Seventy-two minutes of video and audio, compressed at well over a 100:1 ratio, can be stored on a single CD-ROM.[7]

Beyond its compression characteristics, DVI technology and applications have other advantages, including platform independence, speed of operation, and product flexibility.

In the last category, the quality of the final product could vary. Individuals could use a special capture board to process their own information. Or for a higher-quality output, a tape with the pertinent audio and video could be sent to an outside agency for special processing.[8]

Finally, the importance of compression extends beyond the multimedia and desktop video fields. For instance, the development of efficient standards is important for data relays through communications networks. As the communication process increasingly relies on visual information, be it graphics or digitized video, the adoption of standards is critical to the overall enhancement of our communications system. Both developments, advanced communications systems and compression techniques, can go hand in hand.

QuickTime. The overview of data compression leads us to a discussion of QuickTime, a relatively new standard. QuickTime supports applications that allow us to capture and replay digital video and audio on a real-time basis, and the information can be stored on a hard drive or other media. QuickTime serves as a representative example of how this type of system can be used. This includes newer entries in the field, such as Microsoft's Video for Windows.

QuickTime was actually released by Apple in the early 1990s. It brings, in part, digital audio and video to the Macintosh platform at a reasonable cost. Designed to run on most Macintosh computers, QuickTime is an operating system extension.

The latter development is important, both practically and philosophically. Practically, QuickTime made it possible for many Apple users to readily tap into this resource (for example, playback didn't require additional hardware). Philosophically, it made digital video and synchronized audio an integral component of the overall Macintosh system. Multimedia support became a built-in function rather than an afterthought or a hardware and software kludge.

In a typical application, an appropriate video-capture board can be installed in a Macintosh. Connect a videocamera, and you can capture a video clip. After it's saved, the clip and accompanying audio can be manipulated by an editing program.

Figure 8–4

Multiple windows enhance the editing process. The main window, the construction window, shows two clips with the designated transition (placed between the clips). The bottom of the window shows the audio cut. A preview of the final product is shown in the window in the top left-hand corner of the shot. (Software courtesy of Adobe Systems, Inc.; Premiere)

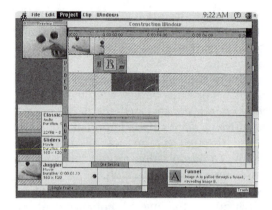

One such package, which illustrates this software category's capabilities, is Adobe Systems' Premiere. The program can join clips together using wipes, dissolves, and other transitions. Special filters can be applied for creative and image-correction procedures, and other effects can be created through various options. The program also accommodates animations, still images, and multiple audio tracks, including the audio recorded with the original video.

Premiere's visual interface makes it a very intuitive editing program. The sequence of clips can be quickly rearranged with a mouse, transitions can be added or deleted, and audio tracks, graphically represented as waveforms, can be manipulated.

Once you complete the project, it can be previewed. If it looks and sounds good, it can be made into a QuickTime movie, where its size and other output characteristics can be controlled.

Depending upon the movie's complexity and length, it may be possible to store the final product on a floppy disk for distribution. The movie can also be used by other Quick-Time-aware or -compatible software. In this case, instead of importing a still image, you import a movie. Place the movie in a document, click on it with a mouse, and the movie plays. Stealing a line from an earlier chapter, it can turn a static document into a moving one.

These characteristics contribute to Quick-Time's value as a communications tool. It can be used with numerous software packages

and, as indicated, does not require special hardware for playback.

Hardware components do, however, have an impact on the system's performance. These range from the computer model itself to the use of a hardware- vs. a software-based compression system.[9]

Yet despite its advantages, there were some shortcomings with the initial Quick-Time release. Images were typically small, and as the movie's size was scaled up, there was a loss of clarity. The frame-acquisition and playback rates could also be low, which could make a movie look jerky, since motions may not have been smoothly reproduced.

Nevertheless, QuickTime and its supporting products are still maturing. Much like the original PCs, they have improved with age and new releases.

It's also important to examine QuickTime, Video for Windows, and other possible standards from a systems approach. When first introduced, a QuickTime movie may have looked great to a computer user. It was cost-effective, and the standard helped extend a PC's production capabilities. A person working with video systems, on the other hand, may have looked at the same movie and wondered what the fuss was all about.[10]

The truth probably lies somewhere between both extremes. When they first entered the market in force in the early 1990s, PC-based digital video standards generally didn't match the quality of conventional high-end video systems. But they were relatively inexpensive to implement, could be accommodated on a network, and could potentially support a higher-quality mode.

More important, PC-based digital systems were and still are flexible. Video can be digitized on a real-time basis, and the information can be immediately manipulated and edited to create a new project.

In one example, you may have to design a computer-based presentation on short notice. To open the project, you may want to incorporate a short digital movie to grab the audience's attention. To complete this work:

1. Use an audio program to digitize your

voice. If applicable, add a special effect, such as a slight echo.

2. Mix your voice with a music clip.

3. Digitize the different video clips.

4. With a program such as Premiere, create an opening title and edit the video clips together with dissolves and other appropriate visual transitions.

5. Save the final combination as a digital movie, and use it as part of the presentation's opening.

The end result may not equal a project created with conventional audio and video equipment. But the PC-based configuration enables you to quickly produce and assemble the various components and, when finished, to seamlessly integrate the movie in the presentation.[11]

What it boils down to is a sense of balance or perspective. Each production environment has its own relative merits and, by extension, its own set of prime application areas. They should all be examined in the context of how they may fit in the overall communications system.

The convergence factor should also be considered. As indicated in different sections of the book, there is a convergence of technologies and applications, including those taking place in the video-editing market.

Chapter 9 describes professional video-editing equipment. A system may allow you to edit compressed digital video and take advantage of the options offered by this environment. When you're finished, the final product is assembled on videotape. This type of system provides you with the best of both worlds: the convenience of *nonlinear editing,* which can speed up and enhance the editing session, and the quality offered by a standard video configuration.

This capability may also be incorporated in preexisting products. An update of Premiere, for instance, added an *edit decision list* (EDL) export feature. An EDL is essentially a list of editing instructions that a high-end editing system can use for auto-assembly, where the master tape, the final program, is assembled from the source material. Important EDL elements include selected transitions and relevant *time code* numbers.

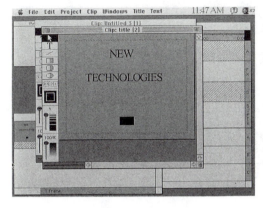

Figure 8–5
Premiere has supported the creation of titles that can be incorporated in a digital movie. (Software courtesy of Adobe Systems, Inc.)

Briefly, time code is an identification process in which video frames are assigned specific reference numbers. A computer-assisted editing system can use this information to identify and select the designated scenes as the final tape is assembled. Consequently, Premiere, originally a QuickTime-editing program, was now capable of interfacing with more advanced systems.

Finally, videotape, like the videodisc, is still regarded as a highly efficient and effective storage medium. While PC-based digital formats are promising, they generally cannot equal, as of this writing, the quality and the storage efficiency of videotape, when matched with high-end video systems.

Graphics Software. Graphics programs play a key role in desktop video and multimedia applications. They can be used to create original pictures, to clean up digitized images, and even to create titles for a show. Pertinent programs were discussed in Chapter 3.

Besides the typical program categories, other graphics-type programs are well suited for desktop video and multimedia projects. CAD programs, for one, can serve as illustration tools. In fact, American Small Business Computers, creators of the DesignCAD software series, discovered that their programs were widely used for tasks beyond CAD applications. Popular choices included business graphics, technical illustration, and desktop publishing.[12] Images generated by a CAD program could also be incorporated in a desktop video project.

Figure 8–6
The Video Toaster's on-screen interface. (Courtesy of NewTek, Inc.)

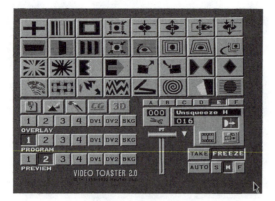

Other pertinent software may be even more specialized. Mannequin, developed for IBM PCs, can be used for ergonomic design applications. But it can also create human figures in different poses. This capability allows designers, and even people who can't draw, to incorporate human figures in their work.

The landscape-generator programs outlined in Chapter 3 can also be valuable tools. For a production, you can generate realistic or surrealistic background scenes ranging from Mt. Saint Helens to the planet Mars.

In a related area, character-generator programs are available for PC products. A character generator is analogous to a typewriter, but in this case, the typed words appear on a television screen. A production's closing credits and the score from a Yankees game are examples of a character generator's output.

Dedicated character generators have existed for a number of years and are still used

by the professional market. But PC owners can take advantage of this production capability, too. Depending upon the software, you may have a choice of typeface, resolution, and text and background color.

Video Toaster. NewTek's Video Toaster is one of those products that spans the professional and nonprofessional markets. Introduced for the Amiga, the Toaster raised the level of desktop video production by at least a notch.

The Video Toaster is an internal board equipped with complementary software. The system functions, in part, as a character generator, frame grabber, and a switcher. Switchers are used for image transitions and are discussed in Chapter 9. The Video Toaster also supports 3-D and 24-bit images, special colorizing effects, and digital video effects where images may spin, flip, or be manipulated in other ways.

The Toaster's importance lies in its delivery of an array of professional features at a low cost and in a single package. But depending upon the production situation, you may also need a *time base corrector* (TBC) to take full advantage of its capabilities. A TBC "takes the unstable video from a VTR and acts as a shock absorber, outputting rock-stable video that can be integrated with other video sources in a system to maintain good picture quality."[13]

TBCs were once very expensive and were out of the financial and possibly even technical reach of most desktop video users. The situation changed, though, when the Toaster's popularity and the growing desktop video market prompted the introduction of inexpensive TBCs, including internal models that fit inside a PC.

Like most PC-based video components, the Toaster's operation is enhanced when used with a faster computer.[14] There are also memory and data-storage guidelines, which when followed will optimize its performance. What you get at the end is a very capable desktop video system, basically a desktop production studio, that can generate sophisticated logos, titles, graphics, and assorted visual effects.

Figure 8–7
LightWave 3D's layout interface screen with real-time wireframe playback. The program, when used with the Video Toaster, can create products ranging from flying logos to graphics. (Courtesy of NewTek, Inc.)

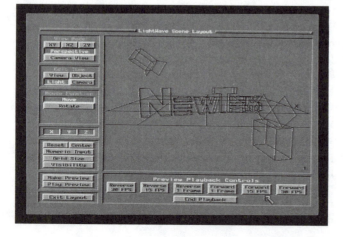

APPLICATIONS AND IMPLICATIONS

Multimedia and desktop video systems have emerged as powerful information tools. Here is a typical application, from the business world: Instead of a business presentation made up of just a series of static slides, a multimedia production can make a presentation more interesting through the use of video and other media. It can support a complete audiovisual environment.

This idea can be extended to teleconferencing, described in Chapter 11, where electronic meetings are held to exchange information. The same capabilities might also reach our homes through high-speed communications lines, as discussed in Chapter 5. Other important applications are covered in the following pages.

Education

The educational market is one of the key areas served by multimedia and desktop video products. As indicated in the hypermedia section, a student can use a multimedia book that incorporates sounds and possibly even video clips. When combined with hypermedia links, the student can explore and experience this new world in a nonlinear fashion. The document's interactive nature, be it created with a hypermedia or multimedia program, can also help make learning active rather than passive.

In the medical field, surgeons could use QuickTime or another digital video product to quickly and inexpensively document new techniques. Besides a written description, a surgical procedure could be annotated with video and voice. This type of product could complement and supplement videotape, videodiscs, and other delivery systems. Digital video additionally offers a rapid turnaround time, an easy learning curve through programs such as Premiere, and the potential to exchange these data over communications networks.[15]

Legal and Broader Implications

Multimedia and desktop video productions

Figure 8–8
Computer graphics, animations, and the law: from a case involving an aircraft accident. (Courtesy of FTI Corp., Annapolis, Md.)

have emerged as tools in the legal field. In a case involving a patent-infringement lawsuit over hip prostheses, a firm created an interactive presentation that depicted how the devices worked. The goal was to show the devices in action and to make the information both accessible and interesting to the jury.[16] The actual production process included using scanned images, image-editing software, and animations.

In addition to full-blown presentations, computer graphics and animations have been used to re-create real-life events. In one case, an animation of a roller coaster depicted the G forces an individual would be subjected to during a ride. The data were gathered by instruments and incorporated in the animation to show how these forces caused a rider to suffer a stroke through a ruptured blood vessel.[17] Other cases have ranged from pinpointing the causes of fires to depicting how automobile and aircraft accidents could have taken place.

Exploring the World. This application, using multimedia and computer animations to explore complex phenomena, also applies to the world at large. As you know, the world is an intricate and rapidly changing system. Volumes of information that cover thousands of topics are generated with each passing hour, and we must cope with this continuous information stream. To make matters worse, much of this information is difficult to comprehend.

Graphics, animations, audio, and video clips can be used to help solve these problems. A desktop video or multimedia presentation may make it possible to distill a mountain of information into a more accessible form. In the case of a report that examines the growth rate of urban centers, a series of graphics could potentially replace pages of census material.

Multimedia and desktop video productions can also be used to explore complicated events. In one example, a series of graphics and animations has been used to explain and depict the Strategic Defense Initiative (SDI), also known as Star Wars. This satellite-based plan, proposed during the Reagan administration, was the core of a very complex and multilayered defense system. The missiles, satellites, and other components, which all played a role in the SDI plan, were brought to life via the computer.

Implications. The graphics and animations made it possible for viewers to grasp, at least fundamentally, how the system would theoretically function. The interactions between the different SDI elements were also depicted in a dynamic fashion and revealed the interrelationships between the component parts. With only a text-based description, this information would not have been conveyed as effectively, and the subject may have been too difficult to comprehend.

But despite this positive attribute, there is an inherent danger in converting a large information base, especially about such controversial and intricate subjects as SDI, into a series of images. The real-world situation may be portrayed in too simplistic a manner. Or the graphics and animations could be manipulated to promote and establish a particular point of view, possibly a distorted view, of the facts.

Since we tend to believe what we see, a production that effectively uses a mixture of media could be misleading. We also generally don't have an opportunity to refute the information presented in a production on a point-by-point basis.

The opposite is true of a courtroom situation when computer-based information is used. In court, expert witnesses could be called to explain, in full, the information portrayed in a presentation. More important, the information could possibly be challenged and rebutted.

Finally, the potential to present an altered view of the real world becomes an even more pressing concern in light of a virtual-reality system's capabilities. A computer-generated world, designed and controlled by a human operator, could serve as an analog for real-world situations. This topic is discussed in the book's Afterword.

Closer to home, images, including news pictures, can be manipulated and altered with a computer, as explored in the next chapter.

Training, Sales, and Advertising

Training applications are particularly well served by desktop video and multimedia systems. A production could cover tasks ranging from car engine repairs to basic PC operations. Video clips of a real engine could be used, and the production could incorporate an interactive interface.

Store owners are similarly served. An electronic sales catalog can either replace or supplement a print version, and interactive kiosks, where customers can get information about products, have popped up in supermarkets and other outlets.

The advertising and public relations industries have also benefited from desktop video and multimedia technologies. Video clips and animations can be rapidly generated, modified, and used in a presentation. In a related area, a sophisticated multimedia system was developed to showcase the city of Atlanta. The presentation was used to promote Atlanta as the site for the 1996 Summer Olympics.[18]

Other Applications

Video artists have also adopted desktop video tools. Inexpensive video systems have been used by artists for years. Earlier projects included personal documentaries as well as video feedback, where different visual patterns, displayed on a television screen, could be created and controlled.

The current generation of artists can now use more sophisticated video equipment and, more pointedly, can tap a PC's power. Video can be digitized for either still or motion displays, images can be colored by computers, and animations can be produced. The affordability and power of computer and video components have provided artists with an array of tools that may have been inaccessible only a few years ago.

Desktop video and multimedia systems also extend the concept of the free flow of information, explored in the previous chapter. As is the case with desktop publishing, a broad user base can now produce assorted electronic documents. Depending upon the situation, these documents may present even more powerful messages than their paper counterparts.

Another important and related point is the capability to create your own production and to place this work in the open marketplace. As mentioned in Chapter 7, you now have another way to express and present your ideas even if no one else views or listens to them.

The growth of the multimedia and desktop video fields could also play a key role in the transformation of the mass media to a personal media. Instead of everyone receiving the same information, more personalized information could be created and received. For example, by using a videocamera and a computer, we literally become producers and editors in our own right.

This concept encompasses the technologies discussed in this and other chapters. Chapter 10, for example, covers electronic information services, including *teletext magazines*. A teletext magazine, an electronic magazine you receive and view on a television set, could relay general news in addition to specialized information. Potentially, a teletext service could offer its subscribers a personalized publication: you would compose your own electronic newspaper or magazine by specifying the types of stories that interest you. Other, more sophisticated electronic information services currently support a similar option.

Various organizations, including MIT's Media Lab, are pushing this concept beyond current boundaries. The Media Lab, directed by Nicholas Negroponte, is one of the world's premier research institutes. In addition to exploring the convergence of different me-

dia, personalized interactive media are also investigated. Two examples are personal electronic newspapers and television.[19]

In this new world, information could be retrieved from different sources and subsequently delivered to you through the assistance of intelligent systems and human—machine interfaces. In one example, a computer could scan a night's worth of programming and then summarize and possibly replay the portions that would be of personal interest to you.[20] Other developments, including an interesting look at the Media Lab in general, can be found in Stewart Brand's book *The Media Lab*.

Aesthetic Considerations

To wrap up this chapter, we should examine some broader, aesthetic issues affecting the multimedia and desktop video fields. Basically, the growth of both markets may have outstripped the development of a sound aesthetic base. An examination of the traditional aesthetic frameworks developed in the film and television industries may prove to be valuable. Multimedia productions serve as an example in this discussion.

Element Integration.   All the media elements that compose a multimedia presenta-

Figure 8–9
PCs, in combination with video equipment, have provided artists with a powerful set of tools. This shot is a still from "Godzilla Hey," by Megan Roberts and Raymond Ghirado (1988). The work combines digital video imagery and sound with analog video synthesis. Produced at the Experimental Television Center with Amiga PCs and the FB-01 video frame buffer with proprietary software, designed by David Jones. (Courtesy of Megan Roberts and Raymond Ghirado)

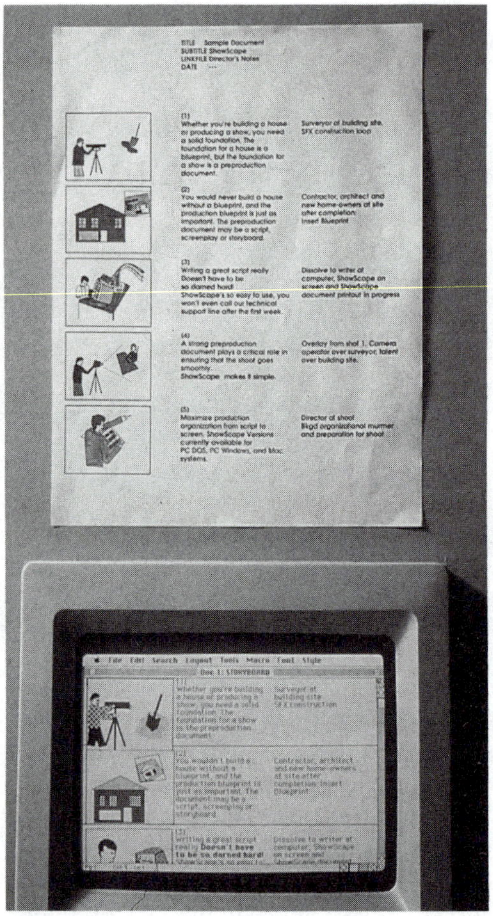

Figure 8–10
PCs, with the proper software, can enhance the writing and production process, in this case, the development of a storyboard. (Courtesy of LAKE Compuframes, Inc., Briarcliff, N.Y.; ShowScape)

During the preproduction phase, different questions, including the following, should be asked:

- What is the presentation's goal?
- Who is the potential audience?
- What are the budgetary limitations?
- What is the best way to satisfy the goal? For example, should video be incorporated in the production?

The last question is particularly important in view of the options you may have in creating a presentation. Many of the tools described in the desktop video section can be used in a multimedia environment. Should you, for instance, use full-motion video or would an animation suffice? Should you use pictures with computer-generated text? Do you need a videodisc player or is digitized video stored on a hard drive the way to go?

Scanners and image-editing software normally employed in desktop-publishing operations could serve a similar function. In this case, a series of scanned pictures may be incorporated in a production. The different graphics programs described in Chapter 3, and those briefly mentioned in this chapter, follow suit.

Other appropriate programs include QuickTime-editing software and audio packages. The flexibility to use such a range of tools is one of the hallmarks of multimedia production. The key again, though, is ensuring the presentation's integrity by using a tool only if there's a reason for using it.

Copyright issues must also be dealt with by the producer. You must be sure that you have clearance to use the sounds and images that compose your production. As indicated at different points in the book, clip art and music are two options, as are original and public-domain materials.

There is also a move toward making it easier for the producer by providing a clearinghouse for licensing arrangements. An organization modeled after ASCAP in the music industry, where copyright fees are collected and disbursed for material usage, could help solve this problem in the multimedia as well as desktop video and publishing markets.[21]

tion should be fully integrated, much like a film or television production. There should be motivation, a reason, for using any given element. An animation should not be used simply because a producer may have the tools to create an animation. Rather, an animation should be used for a specific purpose. It can range from demonstrating the operating principles behind a new piece of equipment to serving as an attention-grabbing device.

Preproduction. This type of issue can be worked out during a *preproduction* phase, the time before the program is actually created. The idea is to establish different criteria to help guide the program's design. The preproduction stage also serves a more practical purpose. It's less expensive to make changes at this stage than after the program's final assembly.

Prototyping. Besides developing a pre-production plan, prototyping a small-scale version of the program may prove valuable, especially if the project is complex. You can try out your ideas with the prototype and plan any necessary changes.

A *storyboard* would also be helpful during this process. A storyboard depicts, in sequential order, the major events in a production. It can be made of still pictures and may include audio. If the storyboard is computer-based, you may be able to extend the storyboard's components and actually use some of them in the final production. These can include graphics, animations, and even QuickTime movies.

Enhancement. Once you create a multimedia presentation, or a desktop video production for that matter, you should seek ways to improve your craft. Continue producing, watch related media products, including movies and television programs, and read.

If you are planning a desktop video project, you can, for instance, learn how music can be an effective component by examining its role in certain movies. Music can heighten the tension in a scene or can serve as a counterpoint to what we're watching on the screen. This principle, observing good movies and television programs for analytical purposes, also applies to lighting, scriptwriting, shooting, editing, and other production techniques.[22]

Finally, the goal of this process should be the development of your own style. Your production, be it video art, an electronic catalog, or a multimedia presentation, can have your own personal signature. With all the tools at your disposal, this should be an enjoyable prospect.

REFERENCES/NOTES

1. Tom Yager, "Practical Desktop Video," *Byte* 17 (April 1990): 108.

2. David A. Harvey, "Local Bus Video," *Computer Shopper*, July 1992, 181. Please note: This concept can be extended to other computer peripherals to similarly speed up their performance.

3. *Microsoft Windows User's Guide* (Redmond, Wash.: Microsoft Corporation, 1992), 451.

4. David Miles Huber, *The MIDI Manual* (Carmel, Ind.: SAMS, 1991), 20. Please see Jeff Burger, "Getting Started with MIDI: A Guide for Beginners," *NewMedia*, November/December 1991, 61, for additional information and for a chart that highlights the MIDI standard's data-storage efficiency when compared with conventional digitized audio.

5. Bruce Fraser, "Scan Handlers," *Publish*, April 1992, 56.

6. Chris Cavigioli, "Image Compression: Spelling Out the Options," *Advanced Imaging* 5 (October 1990): 64.

7. Greg Loveria and Don Kinstler, "Multimedia: DVI Arrives," *Byte* IBM Special Edition (Fall 1990): 107.

8. Britton Peddie, "IBM's ActionMedia II: DVI for the Rest of Us," *NewMedia*, May 1992, 30.

9. Please see Denise Salles and Judith Walthers von Alten, "Chapter 7: Making, Playing, and Sequencing a Movie," in *Adobe Premiere User Guide* (Mountain View, Calif.: Adobe Systems, Inc., 1992), for a comprehensive overview of the latter topic.

10. Ben Calica, "The Clash of the Video and Computer Worlds," *NewMedia*, September/October 1991, 58.

11. Depending upon the project, a digital environment could offer other advantages. With a conventional audio-video setup, an individual may have to learn how to operate a range of equipment versus a few software/hardware packages in a PC-based operation. More important, since you can now see the audio and video information you are manipulating, the learning curve, versus conventional equipment, may be reduced.

12. American Small Business Computers, personal communication.

13. Tedd Jacoby, "Old Problems, New Answers," *Video Systems*, April 1988, 80. Please note: TBCs are also used in high-end, conventional editing systems.

14. The Toaster was originally released for the Amiga. Subsequent plans called for IBM- PC- and Macintosh-controlled products. Through an interface, an IBM or Macintosh computer would control a Toaster in an Amiga. The Toaster's design enabled the manufacturer to enhance its capabilities through a software rather than a hardware upgrade. This capability provided users with additional functions at a reasonable cost.

15. Steve Blank, "Video Image Manipulation with QuickTime and VideoSpigot," *Advanced Imaging* 7 (February 1992): 54.

16. Charles Rubin, "Multimedia on Trial," *New-Media Age,* April 1991, 27. Please note: This case was settled out of court. Opposing lawyers would also have objected to the use of the presentation.

17. Carrie McLean, "Houston Lawyer Makes a Case for Computer Animation in the Courtroom," *Presentation Products* 5 (May 1991): 18.

18. Mike Sinclair, "Interactive Multimedia Pitches Atlanta Olympic Bid," *Advanced Imaging* 5 (March 1990): 38.

19. Nicholas Negroponte, speech delivered at Ithaca College, Ithaca, N.Y., May 29, 1992.

20. Negroponte, speech, May 29, 1992.

21. Paul Karon, "Electronic Publishing Faces Legal Traps Over Copyrights," *InfoWorld* 14 (March 9, 1992): S70.

22. Relevant television programs include "The Twilight Zone" and PBS's documentary *The Civil War.* In cinema, the list runs the gamut from older classics to more modern films: *Battleship Potemkin* and *Alexander Nevsky*; *Grand Illusion; The Adventures of Robin Hood; Citizen Kane; Casablanca; The Third Man; Psycho; The Godfather* (1 and 2); *Raging Bull*; and *Glory.* As a group, the films serve as examples for production and aesthetic principles ranging from music to lighting to shot composition.

ADDITIONAL READINGS

Baran, Nick. "Putting the Squeeze on Graphics." *Byte* 15 (December 1990): 289–94. An overview of different data-compression techniques. Covers still and moving images.

Bernard, Robert. *Practical Videography; Field Systems and Troubleshooting.* Boston: Focal Press, 1990. And, Millerson, Gerald. *The Technique of Television Production.* Boston: Focal Press, 1990. And, Zettl, Herbert. *Television Production Handbook.* Belmont, Calif.: Wadsworth Publishing Company, 1992. Three video-production texts. The topics include, depending upon the book, field production, editing, shot composition, and various aesthetic issues.

Bove, Tony, and Cheryl Rhodes. *Using MacroMind Director.* Carmel, Ind.: Que Corporation, 1990. And, Rosenborg, Victoria. *A Guide to Multimedia.* Carmel, Ind.: New Riders Publishing, 1993. The first book is a review of not only how to use MacroMind Director, a popular multimedia program, but of multimedia presentations in general. The second book focuses on multimedia production (with emphasis on the IBM PC) and includes important production guidelines and a demo disk.

Brand, Stewart. *The Media Lab.* New York: Penguin Books, 1988. As stated in the text, the book provides an interesting look at MIT's Media Lab and its work and implications.

Calica, Ben. "Enterprisewide Multimedia." *InfoWorld* 14 (May 25, 1992): 48–49. Some of the problems facing networked multimedia users.

Cole, Arthur. "Radical Changes Ahead for Video." *TV Technology* 10 (April 1992): 1, 14. A good overview of the convergence between desktop and professional video systems and some of the major issues.

Friedman, Dean. "Sounds of Success." *Byte* 15 (September 1990): 429–42. An excellent overview of MIDI and MIDI systems.

Graves, William H. "Multimedia Manifesto." *Campus Tech* 1 (Spring 1993): 9–13, 29. And, Sherman, Lee. "Digital Video Moves Out of the Trenches." *Presentation Products* 7 (March 1993): 34, 36, 40. Two examinations of PC-based digital video, software/hardware requirements, operations, and for the first article, the multimedia connection.

Huber, David Miles. *The MIDI Manual.* Carmel, Ind.: SAMS, 1991. A detailed guide to MIDI systems and operations.

Lichtman, Andrew. "Forensic Animation." *Amazing Computing* 6 (January 1991): 42–44, 46–47. And, Schroeder, Erica. "3D Studio Gives Crime-Solving a New Twist." *PC Week* 9 (March 9, 1992): 51, 58. These two articles cover other applications of animations in a courtroom.

Luther, Arch C. *Digital Video in the PC Environment.* New York: McGraw-Hill, 1989. An excellent examination of digital video, audio, and the DVI system. The author was one of the DVI system's developers; the book is also available in a second edition.

McDonnell, Ray. "TBCs and Toasters." *AVID.* April 1991, 12–15. An overview of TBCs geared for desktop video applications, especially the Video Toaster.

PC Magazine. March 31, 1992. A good portion of the issue is devoted to the multimedia field. This issue provides a comprehensive look at IBM-PC-based systems and software. Sample articles include: "Authoring Software," by Alfred Poor, 223–49, and "Multimedia," by Michael J. Miller, 112–23.

Wallace, Lou. "Amiga Video: Done to a T." *AmigaWorld.* October 1990, 21–26. An initial look at the Video Toaster.

Wayner, Peter. "Inside QuickTime." *Byte* 16 (December 1991): 189–96. A look at QuickTime.

GLOSSARY

Authoring software: In the context of this chapter, a software classification that can simplify and enhance the creation of a multimedia presentation.

Compression: Compression refers to, for our purposes, reducing the amount of storage space required to store digitized information (for example, video). Compression can also speed up information relays.

Desktop video: Advancements in video technology have made it possible for companies and individuals to assemble cost-effective yet powerful video-production configurations. PCs typically play a major role in this environment.

Digital Video Interactive (DVI): A digital compression system that can support a full-motion, full-screen video display.

Edit decision list (EDL): Essentially a list of editing instructions a high-end editing system can use to assemble the final tape from the original source material.

Musical Instrument Digital Interface (MIDI): A MIDI interface makes it possible to link a variety of electronic musical instruments and computers. The MIDI standard also enables musicians to tap a computer's processing capabilities.

QuickTime: Apple Computer's PC-based digital-media system.

Slow-scan digitizers and frame grabbers: Slow-scan digitizers can capture an image produced by a videocamera (not in real time); frame grabbers can capture an image in real time. In both cases, the information can subsequently be fed to a PC.

Video for Windows: Microsoft's PC-based digital-media system.

9 Computer Technology in the Television and Radio Industries

Besides the revolution taking place in the desktop-publishing and video fields, another revolution is sweeping the television and radio industries. It is a revolution based upon the integration of computers and microprocessors: a new generation of equipment that can enhance a wide range of production and management operations are now available.

As a result of the adoption of computer technology, various equipment functions have been automated. A videocamera, for example, can now be automatically calibrated by an intelligent controller. The processing and production capabilities of graphics generators and other computer-based equipment have likewise been improved.

Television and radio station operations have also benefited from these developments. In two applications, software packages have simplified scriptwriting tasks and have enabled a radio station manager to review a station's advertising and sales records.

There has also been a convergence of technologies, tools, and applications, as outlined in Chapter 8. In the case of video-editing systems, for example, organizations ranging from a television station to a small business can now produce a professional product. In the audio world, digital audio as well as MIDI systems play an analogous role.

Consequently, this chapter examines the integration of computer technology in the television and radio industries. It also covers related developments, including the integration of digital equipment and operations, *high-definition television* (HDTV), which promises to deliver enhanced pictures and sounds to our home, and some of the ethical and legal questions raised by the technologies.

Finally, even though the chapter's focus is on the television and radio industries, it also covers other pertinent and important production environments, ranging from production houses to corporate facilities.

Figure 9–1
One of the fallouts of the communication revolution is the ability to create sophisticated productions with PC-based products. This is a shot from Todd Rundgren's "Change Myself," a music video produced by Rundgren using the Video Toaster. (Courtesy of NewTek, Inc.)

PRODUCTION EQUIPMENT AND APPLICATIONS

Switchers and Cameras

Over the past few years, the *switcher* has benefited from the integration of computer technology. In brief, a switcher is an electronic device that can individually select the pictures produced by a video facility's multiple cameras. These shots can then be recorded on a VTR or routed for transmission.

A switcher can also generate wipes and other visual transitional devices, and it can merge an external video source, such as a news story shot in the field, with a live production, the nightly news program. Thus, the switcher generally controls and manipulates prerecorded pictures stored on videotape and live pictures produced by cameras.

A computer-assisted switcher can help an operator to initiate and complete these picture manipulations. In one application, a complex visual effect with several components may be called for during a production. The specific actions to create the effect can be preprogrammed and stored in memory, so they can be recalled by pressing one button. The capability to immediately execute a command in the middle of a production, when time is always critical, is an important one. A floppy disk can also be used, if the switcher is so equipped, to store the sequence for later use.

Besides this operation, switcher configurations can be stored and later retrieved from a disk. If a television station produces commercials, a nightly news show, and other programs, the switcher may be set up differently to handle the specific requirements of each task. With a stored sequence, the switcher can be rapidly configured for the job at hand.

The influence of computer technology also extends to videocameras. Various operational parameters can be set up with a computer-based control system. This capability can free up an engineer's valuable time and help maintain an important component.

Computer technology has also been used to create robotic camera configurations.

There are certain production situations, such as the taping of a television news program, where the camera shots may vary little on a day-to-day basis. In this environment, it's possible to implement a robotic camera system. Instead of using individual camera operators to frame the shots, this task is completed by the system.

A human operator can actually monitor and control the multiple cameras with various interfaces, including graphics tablets and joysticks. Prestored camera shots and movements can also be recalled, which lends itself to repeatability, and the system may be integrated with a newsroom computer.[1]

While this type of system could function in a range of production situations, there may be some limitations. For example, for sporting events and other dynamic shoots that may demand an instantaneous reaction to capture a specific shot, a robotic system may not equal the speed or capabilities of individual camera operators. There may also be a cost factor, depending upon a system's sophistication.

Digital Special Effects and Graphics Systems

Digital special effects and graphics systems are two of the more powerful tools that have emerged from the marriage between computer and video technologies.

A digital special effects system, also called a *digital video effects generator* (DVE), manipulates a digitized picture or video sequence to create a special effect.[2] The final product may be recorded on videotape or used on a real-time basis in a production.

A DVE can support 2-D and/or 3-D effects; an example of a manipulation is a compressed television picture. After the standard analog video signal is digitized and processed, the original picture can be shrunk or reduced to varying degrees in size and can be positioned on any portion of the screen. The original picture is still visible in its entirety, but now it is physically smaller. Similarly, you can initiate an on-air zoom, in which a title "zooms in" to fill the screen.

Other, more advanced effects include manipulating an image so it appears to flip over,

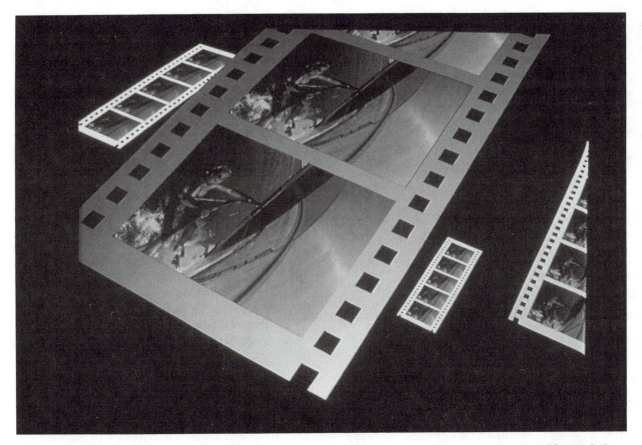

Figure 9–2
A DVE effect. (Courtesy of Pinnacle Systems, Inc., Sunnyvale, Calif.; DVEator™)

much like a page turning in a book. It's also possible to link or wrap an image, or even video, on different shapes.

Since a DVE is a computer-based machine, the instructional sequences used to produce an effect can be permanently stored. As with the system used with a switcher, they can then be recalled to re-create the effect at a later date.

Prior to the late 1980s, DVEs were generally too expensive for production facilities with limited budgets. But the cost for DVEs fell, as did other electronic components of this type, and this tool became more accessible to a broader user group.

PC-based hardware and software configurations accelerated this trend. The price you may have to pay, though, for a PC-based versus a dedicated unit may be speed, flexibility, and image quality. Consequently, both systems are complementary in that they satisfy different, specific needs.

The same developments also influenced computer graphics systems. As previously outlined, a wide range of PC graphics packages has been released. They can complement the dedicated graphics systems on the market.

PC-based combinations can be cost-effective. Another benefit, especially for organizations that use freelance workers, is the available pool size. Since more people are familiar with PC configurations, there is a larger talent pool to tap.[3]

Dedicated systems, for their part, have their own advantages. They can produce and manipulate intricate 3-D images, animated sequences, and digitized pictures. They can also handle large files not easily accommodated by the typical PC-based setup and may have more raw processing power and a broader range of options. These same advantages, by the way, may likewise apply to dedicated character generators, the equipment described in Chapter 8.

Figure 9-3
An on-line editing system with pertinent information displayed on the monitor. (Courtesy of CMX. All rights reserved; CMX OMNI 1000)

Video Editing

Computer technology has also influenced video editing. As described in Chapter 8, editing can encompass the simple joining of scenes or more complex picture transitions, as is the case with A/B roll configurations.

Computer-assisted editing systems have been used for a number of years. They have provided editors with a fine degree of control over the finished product. Two terms are pertinent to our present discussion, *off-line editing*, and *on-line editing*.

Both terms refer to, for our purposes, the editing process. During off-line editing, the sequencing for a program is laid out and can be used to create a "rough cut" of the final program and/or an edit decision list (EDL). The final master tape is then auto-assembled during a more expensive on-line session where a computer-assisted editing console, switcher, and other support equipment that may be required for this operation are available. Thus, off-line editing can save money, since you use a less expensive and complex setup to create the rough cut.

PCs have also played an ongoing role in this process. They have been used during off-line sessions to generate EDLs, which are subsequently input to the on-line system.[4]

But the new generation of PC-based systems has extended their role as editing tools. One of the most widely discussed applications, as of this writing, is PC-based *nonlinear* editing.

Briefly, nonlinear editing allows you to gain access to the different video sequences in a nonlinear fashion, much like retrieving data from a hard drive. In a PC environment, the scenes can, for example, be digitized and then stored on hard drives or erasable optical media. Next, the individual scenes can be quickly retrieved.

A linear system, in contrast, is in keeping with the more traditional editing method. You must search through a videotape to find the specified scenes; this process ultimately eats up more time.

One of the major appeals of the current crop of PC-based systems is the possible marriage between a nonlinear capability and a visual interface, where you see the different audio and visual elements on a monitor as the production is assembled. In one type of configuration, digitized material can be visually represented by video frames, which serve as visual references for the scenes.

Some of the characteristics of working in this editing environment are as follows.

1. With a mouse, the different scenes are selected from an electronic pool or bin and placed in sequential order. You have random access to this information, and a sequence can be quickly assembled, reviewed, and if desired, altered.

2. An audio track(s), such as music, can be added, as can visual transitions between the scenes.

3. A PC-based configuration could generate titles as well as import graphics and animations produced by compatible computer software. In this case, the system could directly tap a PC's graphics capabilities.

4. A digital movie can then be created. While the quality, as of this writing, may not be suitable for high-end applications, the piece could be shown to a client for approval of the final cut.

5. The final tape is assembled, either with the system or through an EDL and a conventional on-line setup. The former capability

has blurred some of the distinctions between on-line and off-line systems. It also gives you the best of both worlds. You gain the flexibility and speed afforded by a digital environment and the potential high quality, the final master tape, produced from the source material.

6. There is a variety of systems on the market, and some setups are cost-effective. Adobe Premiere's QuickTime product, for example, can create an EDL that is compatible with popular on-line editing systems, including the pioneering CMX configurations.

In another example, Digital F/X has offered an off-line Macintosh program. Scenes are digitized with its more expensive hardware module, and multiple users can take this information and work on their own PCs equipped with the off-line software. The final product can then be assembled through the hardware module.[5]

This type of setup can serve multiple users and may be cost-effective to boot. You buy one full-production system and then use relatively inexpensive software for establishing satellite, off-line editing stations.

7. Audio, the sometimes-forgotten but critical production element, can also be manipulated beyond the simple editing of audio clips. Avid Technology, Inc., another PC-based editing manufacturer, developed such a line of tools. Among other features, digital audio tracks could be mixed, and digital effects and the MIDI standard are supported.

Incidentally, the same company developed a digital nonlinear system geared for fast-breaking work, such as news stories. Audio and video could be digitized on a real-time basis, and a story could be edited and directly printed to videotape.[6]

This setup may be representative of one of the new generation of editing systems. Depending upon the application, digital video, rather than an EDL and the original source material, could be used to make the master tape.

8. A patent may play a role in this field. "If a video or film system uses pictures and time code to represent source material, it is using concepts protected by the patent."[7] The upshot is that royalty, licensing, and potentially litigation fees could have an impact in this market's growth, and possibly its future.

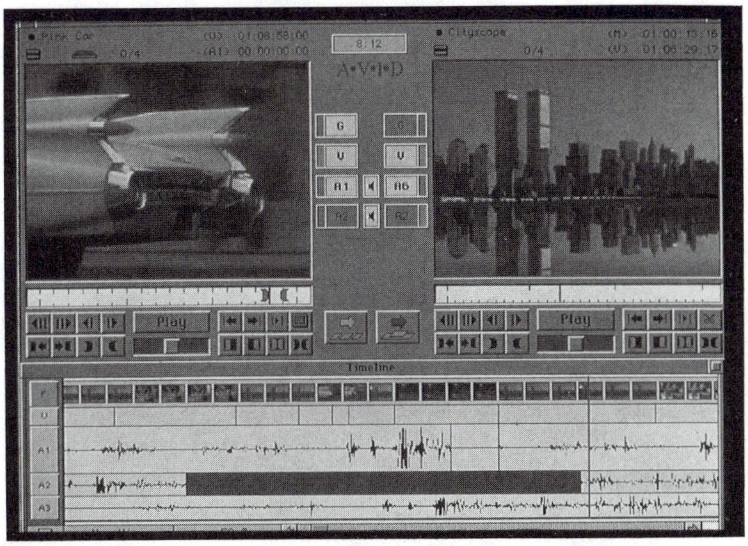

In closing, it's important to remember that PC-based editing is similar, in one respect, to the desktop-publishing and video fields. Beyond the convergence of the technologies that may cut across these areas, all three may take time to master. In the case of editing, while a PC-based system may provide an easy-to-use interface, the actual editing process is still a craft. The system may reduce the learning curve, but it doesn't replace the aesthetic judgment that should accompany an editing session.

Audio Consoles and Editing

Like video equipment, audio components have been influenced by the computer revolution. A computer-assisted audio console, for example, can help an operator to manipulate various sound elements, analogous to the role of its video-switcher counterpart. A system may also support an on-screen representation of a console. To perform various functions, you interact with and control the representation that appears on a monitor, much as you would a physical unit.

An audio signal can also be digitized, and this information can be subsequently edited and manipulated. In the former operation, the process can be performed electronically instead of by physically cutting and splicing

Figure 9–4
The editor's work space: a screen shot of a non-linear editing system. (Courtesy of Avid Technology, Inc.; Media Composer)

Figure 9–5
An example of a digital audio mixer. (Courtesy of SONY Corp.; DMX-E3000)

the audiotape, as is the case with conventional systems. With respect to manipulations, effects can be quickly added, and different variations of the same theme can be saved.

The audio world can also take advantage of some of the features offered by video-editing systems, such as time code information, random-access editing, and the visual representation of a waveform. Some digital audio workstations actually combine recording, editing, and mixing on a single platform.[8]

But, depending upon the application(s), the price you may pay for this enhanced performance and convenience, versus a conventional setup, could be a system's expense. With editing, for instance, you can edit standard tape with a razor blade, grease pencil, splicing tape, editing block, and a tape recorder (for playback and to find the edit points). Digital systems, in this particular case, can't match this setup on a strictly per-cost basis.

Another potential shortcoming may be the necessary information-storage space. Like a video configuration, a digital audio system may have a huge storage appetite.

Finally, see Chapter 8 for an examination of other, relevant audio products, including PC-based editing packages and MIDI systems.

Digital Recording

While previous chapter sections outlined some of the applications and advantages of working in a digital production environment, this section focuses on professional and consumer recording. Specifically, VTRs and DAT systems are covered. Other media, such as optical discs, have been similarly discussed in previous chapters.

Why Digital? Digital audio and video systems have certain advantages over their analog counterparts. Chapter 2 covered some of these characteristics, which include a more robust signal. The recording and storage process also make it possible to preserve a signal's quality even after multiple generations.

This is in contrast to a typical analog system, where in the case of videotape, the output suffers as you progressively "go down" a generation. Digital information can also be readily processed and manipulated, and a digital system may afford an extended recording time.

Audio Recording. The ability to record digital audio has been addressed by *digital audio tape* (DAT) machines. A DAT player can record and play back digital tapes, and the audio quality is equal to that of a CD. A DAT player can pack this enhanced information in a cassette smaller than a conventional audiocassette.

This storage capability made a DAT-type system a viable mass-storage device for the computer industry. More pointedly for this chapter, DAT's recording and operational characteristics made it particularly attractive to professionals. Studio and field machines have been designed, time code can be supported, and the output is excellent.

The consumer market, though, has been a different story. On face value, DAT systems should have been a success. Yet various factors combined to create, at least in the United States, a flat consumer response.

One of the problems was a concern on the part of the recording industry for the home-taping market. For example, companies claimed consumers would make direct CD-to-DAT copies, that is, digital-to-digital recordings, thus potentially reducing CD sales.

Various protection schemes were proposed to deal with this situation, including the Copycode system. In practice, a DAT machine would shut down if it detected a special "notch" in prerecorded media. But this system was dropped since it didn't work all the time, and some individuals indicated it affected the playback quality of the media it was supposed to protect.[9] When the confusion brought about by this situation was combined with other factors, such as the threat of litigation against manufacturers who sold unmodified DAT machines, the result was the flat market.

While this picture has somewhat improved, DAT equipment geared for consumers may actually end up as a niche market for audiophiles, unless prerecorded tapes become widely available at competitive prices.[10] An-

other factor is the potential emergence of other digital recording media.

Two such products are the *digital compact cassette* (DCC) and a recordable mini-CD. The DCC was a digital cassette system, introduced by Philips Electronics, that was backwards compatible. Basically, a player could accommodate the new DCC format and analog audiocassettes. The system also featured a high-quality audio output and a more rugged cassette design.

The second product, Sony's *Mini Disc* (MD), was a portable CD recording system. It combined a random-access capability with digital-quality sound. It also had technical enhancements, including a system that could compensate for physical jarring, allowing the playback to be undisturbed.

Despite their positive features, the DCC and MD had some critics. The DCC, for instance, used a digital compression scheme, and some experts claimed it did not equal a CD's sound quality.[11] The MD, for its part,

Figure 9–6
An example of a digital videocassette recorder. (Courtesy of SONY Corp.; DVR-20)

when first announced, was not backwards compatible with standard CDs.

On a more positive note, both companies could tap into a broad and deep pool of prerecorded works. A pirating protection scheme, which was designed to prevent mass copying but would allow a consumer to make a high-quality personal copy, was also available.

Besides supporting such a protection system, there have been legislative moves, such as the Audio Home Recording Act, to give musicians and publishers royalties on the sale of blank cassettes. Both actions would defuse some of the problems that originally affected the DAT upon its release.

Finally, it's appropriate to end this section with a discussion about the *cart machine*, an audio component that is a common fixture in most professional facilities. This device employs a cartridge, the cart, with a variable storage capacity, and more recent technical enhancements have included the use of microprocessors for monitoring and control functions.

Digital replacement systems have also appeared on the market. In one configuration, audio can be stored on removable magnetic media. It is also possible to build an automated and integrated system. Instead of employing individual carts or their replacements, the audio cuts are automatically retrieved from hard or optical drives, based upon an operator's instructions.[12]

Video Recording. As indicated at the top of this section, one of the advantages of a digital VTR is its high-quality, multigenerational capability. The competing formats include *D-1* and *D-2*.

The D-1 system is a component versus a composite digital standard, and for our discussion, this has implications for a system's quality, applications, and compatibility within an existing production environment. In essence,

the D-1 system . . . is especially useful if you plan on extensive post-production that involves many tape generations . . . and extensive special effects. By keeping the R,G,B components separate, the image quality remains basically

unaffected by even the most complex manipulation. . . .

The D-2 system, on the other hand, does not require any modification or replacement of the existing equipment. Because the D-2 VTR processes the video signal in its composite NTSC configuration, you can use it instead of regular . . . VTRs and in tandem with regular switchers and monitors.[13]

Consequently, a component system can provide a higher quality product, especially in certain production situations. The D-2 system, in contrast, is generally less expensive and, as described by Herbert Zettl, is compatible with the existing physical plant.

The digital VTR market is not limited, though, to these formats. Much like the analog market, the digital end of this industry may be similarly populated. Existing standards, as of this writing, also include the D-3 format, which has particular field applications, and more are on the way.[14]

The latter also encompasses disc-based systems. They may not match a VTR's storage capabilities but could speed up the retrieval and playback process.

INFORMATION MANAGEMENT AND OPERATIONS

Besides the integration of digital and computer-based production equipment in the physical plant, another development has played a major role in the broadcasting and nonbroadcasting industries. Computers have been adopted for numerous information-management and -control operations. These include scriptwriting, newsroom automation, and computerizing a station's traffic and sales departments. Even though these tasks may not be as visible as the production end of a facility, they are nonetheless vital.

Scriptwriting and Budgets

One of the critical jobs in any production facility is scriptwriting. The script is the heart of any program, and as such, software has been developed to enhance this task.

In general, a scriptwriting package frees the writer from time-consuming mechanical chores, such as numbering scenes and adhering to the correct format for margins, columns, and headings. The software completes these functions, and the writer can concentrate on the precise and creative manipulation of the written word.

A program may support various styles, such as a standard two-column television script with video and audio columns. Another valuable feature may be the linking of specific sections of these columns. If changes are made, the corresponding audio or video section tags along.

This type of dynamic adjustment cannot be matched with the typical word-processing program. Even if multiple columns are supported, you may have to continually make spacing adjustments as you work.

The scriptwriting process can be further enhanced through supporting software. In one instance, dedicated and nondedicated programs can be used to create a storyboard, essentially a road map of a production. A storyboard can help you to visualize the final product, and it may even be possible to use a Premiere-type program in this operation. Assemble short QuickTime movies, place them on a page, and then activate them with a click to follow the story's progression.

Finally, other specialized and general-release programs are widely used in the industry. For budgets, spreadsheets can track business expenses as well as crew and talent fees, travel and post-production costs, and equipment rentals. You can also monitor these expenses as bills and invoices are received, and actual costs can then be compared to the projected budget. Another option includes the use of CAD programs for facility design.

News

Computer technology has played an instrumental role in transforming the traditional news department into a highly sophisticated electronic newsroom. Newsroom packages have typically been configured either as a software-only or as a turnkey system. The first option is designed for organizations that

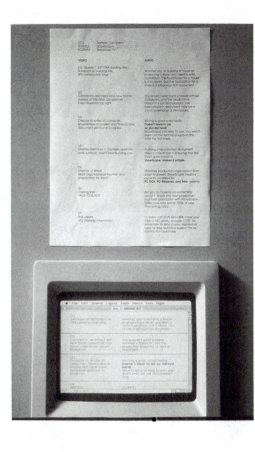

Figure 9–7
PC software can simplify the screenwriting process. In this case, a two-column format is displayed. (Courtesy of LAKE Compuframes, Inc., Briarcliff, N.Y.; ShowScape)

may already own a computer; the second would include both the software and possibly a networked computer system.

A newsroom package, regardless of its configuration, may encompass a number of modules that can support different activities. The news stories, for example, can be written with a newswriting/editing module, which incorporates functions beyond the traditional word-processing program. These may include a split-screen feature where you can write a story while simultaneously viewing the pertinent wire story, all on the same monitor.

A wire-service module, in turn, can supply reporters, producers, and other newsroom personnel with direct access to wire-service copy that can be fed into the newsroom's computer. At this point, a keyword search can be conducted to produce screen summaries of the various stories carried by the service, and then the full text of selected stories can be retrieved.

Basically, this type of system marries traditional wire-service capabilities with the computer. Information is delivered in a timely fashion, and the computer can function as an information-management tool. The wire copy can be retrieved when writing a story, hard copies can be printed, and the information can be organized.

The newsroom package may also include an archiving function. This module, an electronic news morgue, can be used to save and catalog current and past stories, assignments, and other data. This information can be retrieved in a matter of seconds, and the module may incorporate a security mechanism to prevent unauthorized personnel from gaining access to the database.[15]

A networked operation may subsequently enhance the distribution of information in the newsroom since an editor could gain access to the stories written by a reporter tied into the system. This same integrated configuration could facilitate the flow of information throughout the organization, via E-mail, and could make other resources readily available. These include graphics and potentially digital video clips. As compression techniques improve, a news morgue could contain video clips, both for editing and reference purposes, besides text-based information.

Finally, depending on the manufacturer and package, other operations could be supported. These include a newsroom-management function, such as an electronic assignment desk, as well as an interface to the station's production facilities.

In the latter situation, the newscast's text could be seamlessly fed to a *prompter* and to a *closed-captioning* setup. A prompter is a device used by on-air talent to maintain good eye contact with a camera while reading news copy. Closed captions are the normally invisible subtitles for programming that can be displayed on a television set through a special decoder. Closed-captioning can serve a hearing-impaired audience.

Consequently, newsroom packages can accommodate a range of tasks. A product can be used alone, as may be the case with a newswire-service feed to a single PC. Or a comprehensive and integrated system that links the newsroom with the production end of the operation can be created. In fact, Basys Automation Systems has extended this principle to tie character generators, video cart machines, robotic camera systems, and still stores in a central control network; this concept could be further extended to the entire facility.[16]

Operations and Information Management

Research. Broadcasting and nonbroadcasting organizations, such as cable companies, have adopted PCs and minicomputers for marketing research. Arbitron, for one, has been a major force in this field and has offered numerous services for the television, radio, and cable industries. In one example, ratings information can be downloaded or retrieved from Arbitron's computer. This capability provides almost immediate access to valuable information without the waiting period normally associated with hard-copy or disk-based releases.

Arbitron has additionally offered software tools for analytical purposes: you can now draw more information from the raw facts. If appropriate, charts and graphs geared for sales applications can also be generated.

In the course of its work, Arbitron has

Figure 9–8
Computers have played a major role in transforming the newsroom, including the newswriting process. In this application, a split-screen display shows script preparation with imbedded machine-control commands (top) and the wire-service display (bottom). (Courtesy of Basys Automation Systems)

employed various audience-measuring techniques. In one case, viewers have responded to on-screen prompts as part of an interactive television-measuring system. This operation has provided the company with more accurate information about audience viewing preferences than may be afforded by a more traditional method, in which people write down this information in a special diary. Arbitron has even investigated an AI-based passive system that could identify different viewers without on-screen prompting.[17]

Besides these operations, software has been released to support telephone surveys. In brief, questionnaires can be designed and data can be collected and stored on the computer either in real time, as a telephone survey is actually conducted, or after the survey is completed. The data are later analyzed to reveal the characteristics of the viewing or listening audience. In one typical application, these new insights can be used to target and attract a specific demographic group.

Before the PC's widespread adoption, the software and hardware to initiate a survey were typically beyond the means and technical expertise of the average broadcaster, and a third party had to be contracted to complete this task. The newer PCs and the complementary programs have somewhat altered this picture, and it may be feasible for even small-market stations to start a regular survey schedule.

Programming, Traffic, and Sales.   Software can also be used as an organizational and managerial tool. A computer program can generate a detailed list of a day's programming events and can handle other tasks. A radio station serves as an example.

In one application, a program can support a music library, essentially a sophisticated database. Each song in the library can be cross-referenced by artist and title name and, possibly, an assigned cart number. Thus, a song can be accessed in more than one way.

Songs can also be coded according to their tempo, intensity, or any prescribed restrictions, such as an instruction not to run a specific song during the morning drive time. The software may also link age-group and demographic-appeal codes with each song. These data are useful in attracting specific audiences. For instance, if a broadcaster is interested in the 18–34 age bracket, songs can be selected that have been coded to match these demographic characteristics. In a similar vein, the software may help a program director to devise the music-rotation schedule, the list of songs played during the day.

Programs have also accommodated traffic duties. Log maintenance and the subsequent production of reports are essential tasks for any station, and computers have sped up this process.

A program could also automate the scheduling of commercials, the spots. Computers have made it possible to schedule spots more than a year in advance and to provide time separation for competing clients: one competitor's commercial would not immediately follow another competitor's.

In a related application, sales-force performance could be evaluated in different categories, including the dollar amount of sales, the number of sold spots, and each salesperson's commissions.

An accounting package may also be used to complete an integrated system that could cover everything from music-rotation schedules to commissions. Accounting software performs a multitude of billing, payroll, general ledger, and projection functions. In sum,

Figure 9–9
An HDTV display. Note the size and shape of the screen in comparison to the typical TV in current use. (Courtesy of the Advanced Television Research Consortium; AD-HDTV)

the accounting software could handle most, if not all, of a station's financial and book-keeping needs.

HIGH-DEFINITION TELEVISION

The first part of the chapter presented an overview of some of the computer-based developments and digital applications that are shaping the television, radio, and non-broadcasting media. This section focuses on HDTV, an application designed to produce an enhanced television display. The section concludes with an overview of an analogous system geared for the radio industry, *digital audio broadcasting* (DAB) operations.

Introduction

HDTV is an application designed to produce a display that appears to rival the quality of 35mm film, even though it may not, techni-cally speaking, equal film's resolution. An HDTV relay will also support an enhanced audio signal and a 16:9 aspect ratio, that is, the ratio between a screen's width and height. In contrast to the standard 4:3 con-figuration, an HDTV screen would have a greater visual impact, thus presenting a more powerful image to viewers.

The impetus for the development of HDTV and other advanced systems, which can fit under the *advanced television* (ATV) designa-tion, stems from a natural process: the im-provement of the current level of technology and the growth and change that takes place in any industry.[18] The trend toward larger television screens has also accelerated this research, since in general, as the screen size increases, the picture quality decreases. But HDTV and other ATV operations could im-prove upon current standards by providing enhanced displays.

Background

NHK and CBS. The Japan Broadcasting Corporation, NHK, has been one of the world leaders in the HDTV field, and its personnel have conducted research in this area for a number of years. This includes experiments to test a range of technical and nontechnical parameters, such as the relationship between the number of lines in a picture, the opti-mum viewing distance from a screen, and the screen's size. For example, viewers judged the quality of television pictures com-posed of different lines of resolution from various distances.[19] These considerations were weighed, in turn, against bandwidth limitations, the number of lines that could be accommodated in any given system, and other factors.

Beyond this experimental work, NHK conducted trial HDTV transmissions in its own country, including satellite relays as well as demonstrations in the United States. NHK also developed the equipment required to operate an HDTV system.

CBS, for its part, was one of the early supporters for an HDTV system in the United States. The company promoted a two-channel, 1,050-line component scheme, in contrast to NHK's 1,125 configuration. The CBS operation was designed for a DBS relay, and a 525-line signal and enhanced picture information would have been carried in two separate channels. A receiver would have subsequently combined both channels to produce a 5:3 HDTV display.[20] For viewers without HDTV receivers, the single 525-line channel would have been converted for viewing on conventional sets.

HDTV Developments

The work conducted by NHK, CBS, and oth-ers helped spur the development of HDTV technology. As of this writing, HDTV relays are on the verge of becoming a reality in the United States.

This situation has been an evolutionary one, with a minor revolution thrown in to boot. The evolutionary phase encompassed what appeared to have been an emerging infrastructure to support HDTV productions and relays. The revolution was the possibility of developing an all-digital system.

Global Issues.

He who controls the spice, controls the universe.
    —From the movie version of the book *Dune*

During the 1980s, various attempts were made to create a worldwide HDTV production standard, which can be viewed as a different issue from the transmission standard. While there was some progress in the overall HDTV standards process, the only substantial official consensus came when countries agreed to disagree for various technical, political, and economic reasons.

Part of the problem was the concern that one country, Japan, would gain an economic and technical edge in this field if its HDTV system was adopted as the international standard. By extension, the situation could have had an impact on sales in the consumer electronics industry.

In the case of Europe, manufacturers believed that Japan could "swamp" their home markets.[21] Other issues stemmed from the European Community's goal to promote European technologies and, ultimately, European programming.

In the United States, there was initial official support for a standard based upon Japan's system. But this support faded as the prospect for a global standard faded.[22]

Another factor was the state of the American semiconductor industry. To some, HDTV represented this industry's future, since HDTV systems would be heavily tied to semiconductor technology.[23] It was claimed that if the United States was not a leading HDTV manufacturer, the country's overall semiconductor industry could suffer and, by extension, dependent markets (for example, PCs).

Accordingly, a consensus was not reached for an HDTV standard on an international level. It was feared that the country that dominated this industry would have a hammerlock on a multibillion-dollar business and related industries.

This concern and perception is best summed up by the line that opened this section, from *Dune*, the classic Frank Herbert science-fiction work: "He who controls the spice, controls the universe." Only in this case, the spice was and is HDTV.[24]

United States.   While the attempt was made to establish standards, HDTV developments kept pace. In Japan, work continued on its MUSE system, while the Europeans launched a new initiative in the mid-1980s, the Eureka Project.[25] Both systems were geared for satellite delivery.

In contrast, much of the focus and regulatory maneuvering in the United States has been on the terrestrial system, over-the-air broadcasts. This is a reflection of the system's unique status. Unlike cable or another optional service, terrestrial broadcasts were and still are free. Even though the industry is supported by commercials, consumers don't pay a set fee to view the programming.

The FCC helped accelerate the development of such terrestrial HDTV operations in the late 1980s and early 1990s. In a 1988 decision, for example, the agency supported, at least for terrestrial transmissions, an HDTV configuration that would essentially conform to existing channel allocations and would be backwards compatible, possibly through *simulcasting* or an *augmentation* channel. Both terms are discussed in later paragraphs.

Another issue for any HDTV service is the channel-space requirement. An HDTV relay packs in more information than its conventional counterpart, and for the United States, it would exceed, under normal circumstances, the current 6 MHz television-channel allotments. Thus, various organizations attempted to develop a system that operated within these constraints.

In one example, the David Sarnoff Research Center supported its upgradable configuration, the *Advanced Compatible Television* (ACTV) system. Under ACTV-I, an enhanced picture, but one not equal to what has commonly been viewed as the 1,125-line HDTV standard, could be relayed over existing channels. At some future date, an augmentation channel, a second channel with the enhanced information, could be integrated in the system through ACTV-II, to produce a higher-quality picture.

An ACTV-I configuration, also known as *extended definition television*, would have functioned as a bridge between conventional television and HDTV. It

• was backwards compatible;

- accommodated an enhanced picture or display with an appropriate set;
- didn't disrupt current spectrum allocations; and
- could handle a true HDTV relay through the augmentation approach.[26]

Other related issues have included the differences between the broadcasting and the cable and satellite industries. Limited spectrum allocations translated into tight channel requirements for broadcasters. Cable and satellite operators have more flexibility in this regard and could launch, as another option, HDTV pay services for consumers who opted for this programming.

Evolution and Revolution. The FCC further refined its decision in the early 1990s by announcing its support for simulcasting. In this operation, a station would continue to relay a standard signal. The station would then be assigned a second channel for a compressed HDTV relay.

This approach to delivering HDTV was viewed by some as having an important advantage over augmentation. With simulcasting, the HDTV transmission was not tied to the NTSC standard. Thus, a superior, non-compatible system could be developed.[27]

The final piece to the puzzle was added when the FCC, led by former chairman Alfred Sikes, encouraged the development of a digital system. If successful, this configuration would have certain advantages over analog operations.

Similarly, consideration was given to a flexible standard that could handle future growth. Three terms were associated with this philosophy, *scalability, extensibility,* and *interoperability.*

In a scalable video system, the aspect ratio, number of frames per second and the number of scan lines can be adjusted . . . to the requirements of the individual picture . . . or viewer's choice. Extensibility means that a new television system must be able to operate on diverse display technologies . . . and be adaptable for use with new, higher resolution displays developed in the future. Interoperability means a television system can function at any frame rate on a variety of display devices.[28]

Basically, part of the HDTV deliberation process was the consideration of a standard that could accommodate a range of configu-

Figure 9–10
Engineers set up the encoding room at the foot of the transmission tower at WRC-TV/ Washington in preparation for the first over-the-air digital HDTV simulcast, in September 1992. (Courtesy of the Advanced Television Research Consortium; AD-HDTV)

rations and could support future enhancements.

In preparation for the eventual standard, several proposed terrestrial systems, which included an analog system, were scheduled for testing by the Advanced Television Test Center. Following the tests, the FCC was slated to select a standard in early 1993, which was subsequently moved to a later date.

Besides these developments, the FCC worked on a timetable for the implementation of terrestrial HDTV broadcasts in the United States. This encompassed items such as the length of time NTSC signals would be supported before a full switchover was made to HDTV.

All broadcasters have not been happy, though, with these decisions. One factor, especially for an industry affected by the slow economic growth of the early 1990s, would be the financial cost. Estimates for this conversion have ranged from $10 billion for broadcasters to an even higher figure for consumers.[29]

Since HDTV would be a new ballgame, both producing organizations and consumers would have to buy new equipment to play. The cost could be particularly painful for small-market stations, even with eventual price reductions as equipment became more widely available.

Other broadcasters were dissatisfied with the FCC's timetable and HDTV mandate. It was proposed that the second allocation could be used for digital multichannel television relays, or possibly for interactive and/or data services. The goal was to provide broadcasters with additional programming options so they could better compete with cable and potentially satellite services. Eventually, HDTV relays could be phased in, or as it was suggested, the advanced services could be retained since over-the-air HDTV broadcasts may never actually be initiated in the United States.[30]

Other Considerations. The adoption and implementation of an HDTV standard may alter the way programs are produced and the way we use television. Some of the applications and implications, as well as questions that remain about this service, include the following.

1. Even though HDTV is not yet an established global broadcasting tool, it has been and is being used in the United States and other countries for various applications and shoots. In one case, Japan has pioneered the integration of HDTV as a part of everyday television programming.

2. As of this writing, manufacturers have released equipment that would conform to the new, wide-screen format and could provide an upgrade path to HDTV operations. Although the sets are not true HDTV units, they could show a movie without *letterboxing*. (Letterboxing is the technique used with conventional television to display a theatrical movie as it was originally shot, without the cropping that may occur during a typical broadcast. But it mandates the use of black bands running across the screen's top and bottom to accommodate this view.)

With a newer set, the image would fill the screen if a movie stored on a videodisc, for example, supported this format. Camcorders that could operate in either viewing mode have also been developed.

While these sets could provide an enhanced viewing experience, there is a problem when displaying conventional programming: due to the set's format, letterboxing may have to be employed. This same scenario could also apply to HDTV during the transition period from an NTSC to an HDTV standard.

Even though various techniques could be used to handle the "aspect ratio incompatibility," viewers may run into other problems.[31] The picture quality could suffer, among other image and viewing degradations.

3. Beyond the broadcasting community, other major players in the HDTV field could include the satellite, cable, computer, and film industries. Consequently, it has been proposed that standards shouldn't reflect only broadcasting interests.

In the case of the film industry, HDTV could be an attractive production medium. Yet to compete with 35-mm film, the HDTV standard should be flexible enough to accommodate a higher-resolution image than the one afforded by the projected systems.[32]

The flexibility point is important. As described throughout the book, one of the hallmarks of the communication revolution is, in fact, flexibility. New communications channels can accommodate a range of information and support multiple applications. For HDTV, if the system is too narrowly defined, it may not be able to incorporate new technological developments and handle additional applications.

In contrast with this viewpoint, we may have to balance the needs of the future with the concrete concerns of the present. For example, how do we define a standard that can serve broadcasters right now but may have the flexibility to support the computer and film industries? Apparently, by some of its deliberations, the FCC believed both needs could possibly be accommodated.

4. As indicated, while the HDTV focus in the United States has been on terrestrial broadcasting, other media could deliver HDTV programming and specialized ATV services on an optional basis and, perhaps, on a more flexible timetable. Broadcasters, though, are faced with a double bind.

A station could receive an extra channel to relay HDTV programs. Yet the station must abide by an FCC timetable concerning HDTV program implementation and, as of this writing, may have to surrender a channel when HDTV broadcasting is in full swing.[33] Thus, unlike cable and satellite services, a broadcaster has limited choices when it comes to HDTV and may still be left with only one channel after a multimillion-dollar expenditure.

5. An HDTV system could have numerous nontelevision programming applications. An HDTV display could, for instance, be particularly attractive for organizations engaged in videoconferencing activities, the electronic meetings described in Chapter 11. Special educational programming that may highlight delicate surgical procedures could also be produced, and the Defense Department has similarly expressed an interest in this field.

Consumers could be served as well. A set could support various information services and an enhanced display for videotape and optical-disc systems, two other potential HDTV delivery vehicles. In this environment, the set could be the core of an advanced home-entertainment and information center.

6. HDTV transmissions could help promote another entertainment form. A satellite could deliver HDTV signals to special movie theaters equipped with large, high-resolution screens. The programming could include theatrical movies transferred to videotape prior to their uplink, movies shot in an HDTV format, and live as well as recorded concerts and special sporting events. But, as indicated in point 3, a higher-resolution system may be required for these applications.

7. A central question, and one that is still unanswered, is the consumer's role. Does the average consumer want HDTV? More pointedly, is the average consumer willing to pay a substantially higher price for HDTV sets?

These questions are important ones, especially in a competitive consumer electronics market where different technologies and applications are competing for the consumer dollar. Other factors include the state of the economy and the ability of program producers and distributors to sustain the HDTV market until a critical viewing mass is reached.

Conclusion

As of this writing, the HDTV field is still in a fluid state. It also appears it will remain unsettled, at least in the United States, through the early 1990s and possibly beyond this point.

This situation is a reflection of the variables that have and will continue to play a role in the field's development. The players range from the FCC to broadcasters to the consumer, and although you can try to predict what the future might bring, there are a number of alternate scenarios.

For example, will the adoption of HDTV components follow the route of the CD as outlined in Chapter 2, or will consumers reject HDTV at this time? The latter option, market rejection, had a detrimental impact on consumer-oriented videotex services described in Chapter 10, and on other technologies and applications.

If a similar situation arises with HDTV, how do you respond? Do you extend the FCC's timetable for the implementation of an all-HDTV terrestrial relay? Or should you adopt an ACTV-I-type system in the interim, give broadcasters more flexibility in programming their channels, and slowly phase in HDTV? Another option is to promote *improved definition television* techniques, which could include receiver enhancements, that would work within the current NTSC infrastructure to potentially produce a higher-quality display. You could also focus on narrowcasting, possibly through cable or satellite, where you may be satisfied with reaching a smaller audience as more people gradually buy HDTV sets.

In contrast to these scenarios, HDTV could prove to be very popular, as some market projections have suggested.[34] We may eventually witness terrestrial, satellite, and cable relays, or at the very least, some sort of ATV support by all three. Videodisc and videotape equipment may also play a role in these developments, and the overall field was given a boost in the mid-1990s. Competing manufacturers pooled their resources to develop a single standard.

Finally, in closing, HDTV isn't the only broadcasting operation that promises to deliver an enhanced product. Digital technology may also influence radio programming through DAB operations.

DIGITAL AUDIO BROADCASTING

Pioneered in Europe, DAB could deliver a CD-quality broadcast to listeners. The relay could be by terrestrial means or by satellite.

A frequency allocation (L-band), granted during a 1992 World Administrative Radio Conference, initially received wide support by countries pursuing a DAB option. But the United States backed another plan since this allocation was already used for other services.

The National Association of Broadcasters

Figure 9–11
Side-by-side comparison: NTSC on the left and AD-HDTV on the right. Note the aspect ratio and screen size differences between the two. The HDTV system also supports a much higher-resolution display. (Courtesy of the Advanced Television Research Consortium; AD-HDTV)

(NAB), for its part, has been cautious in its approach toward the technology. As of this writing, the NAB has "recommended that any domestic inauguration of DAB be on a terrestrial only basis, with existing broadcasters given first opportunity to employ the technology."[35]

This recommendation stems from two related concerns. The first is that a satellite-delivered service could adversely affect the broadcasting industry, much like DBS television relays discussed in Chapter 6 could have an impact on broadcasters.

The second concern is that of localism.[36] Basically, a satellite-based DAB operation may not match a conventional radio station's public-interest commitment to the local community. The public may not be as well served.

In contrast with these viewpoints, satellite DAB services, which could potentially use another allocation, may have their own benefits. National programming could, for example, supplement local radio broadcasts. It may also be possible to launch pay services, much like one of the HDTV options, and support narrowcasting. A complementary terrestrial and satellite system could likewise develop.

Besides these possibilities, other factors could have an impact on the industry. This includes work on in-band systems.

As envisioned, special techniques could

make it possible to relay programming in existing allocations. This configuration may accommodate, at least in terms of spectrum space, the current radio broadcasting industry. It could also provide stations with an upgrade path to digital relays.[37]

Consequently, while DAB systems are promising, the industry, as of the early 1990s, was still in a state of flux. There was also, as always, the question of consumer acceptance.

OTHER CONSIDERATIONS

Colorization

To close this chapter, and in following the organizational scheme adopted for other chapters in the book, it's appropriate to discuss some of the ethical and legal questions raised by the integration of the new technologies. One such computer application in the film industry, which likewise has implications for the television industry, is the *colorization* of black-and-white movies.

A powerful computer is mated with sophisticated computer software to manipulate information, a black-and-white movie, to create a new piece of work, a color version of the original film. Black-and-white television programs have similarly been altered to generate new, colorized versions of the original shows.

In brief, when NASA started the exploration of the moon, Mars, and other planetary bodies in the solar system with outer space probes, thousands of photographs of these worlds were transmitted back to Earth. The photographs were later converted into a digital format, if they were not already in this form, and were fed to a computer. The picture information, the data that composed the photographs, were then manipulated in a fashion analogous to the way a computer would manipulate other digital information.

This operation, called *image processing*, had two primary goals. The first was to correct picture defects; the second was to enhance various characteristics of the original pictures. For example, if a picture was marred by noise and other defects, they could either

be eliminated or minimized to produce a technically superior picture. Enhancement techniques, on the other hand, were used to manipulate the picture information to help an individual to better interpret the photographic data. In one case, a photograph's contrast could be altered to highlight specific physical characteristics of a region of Mars.

When NASA initiated this program, the state of the art dictated the use of computer systems that only a select group of organizations could afford. But a new generation of cost-effective yet powerful computers, matched by sophisticated software, has made image processing a viable tool in many fields.

Physicians, for instance, have processed X rays to reveal details of a patient's body that were invisible in the original images; astronomers, among others, have adopted a technique called *pseudocoloring*. In this operation, a black-and-white picture is converted into a false-color image.

As you may recall, a black-and-white picture is composed of gray shades that correspond to the variable brightness range or levels in the original scene. If several areas in a galaxy or other photographed object are very similar in this respect, they may be reproduced as almost identical shades of gray. It may be impossible to visually distinguish between the shades and, thus, the galaxy's physical characteristics.

Pseudocoloring can improve this situation. By using the computer, specific colors are assigned to the different shades of gray. The picture information is then manipulated, and all the relevant shades of gray are reproduced in the appropriate colors. This new photograph will now clearly highlight the physical features of the galaxy since they are represented by different and contrasting colors rather than the similar gray shades.

These various image-processing techniques are the basis for many of the picture-manipulation procedures used in our own industry, the communications field. While the actual process and the final product or result may be different, the intent is essentially the same: to manipulate a picture for a specific purpose, be it for a scientific application or for a desktop-publishing project. In one specific example, image-manipulation techniques have contributed to the colorization process.

In the mid- to late 1980s, the colorization technique was applied to black-and-white films. Movies such as *It's a Wonderful Life* and *Yankee Doodle Dandy* were processed by a computer, and colorized versions of the films were produced, to the cheers and jeers of supporters and opponents alike.

During this process, a videotape copy of the picture is created, and a number of operations are initiated to add color to the film. One important element of this multifaceted task is the assignment of specific colors, which correspond to those used when the movie was produced, to the appropriate objects in a frame. A computer then processes the film, frame by frame, and a colorized version of the original movie is eventually generated. In the final product, a former gray hat and suit may now emerge as a bright yellow hat and a sky-blue jacket and pair of pants.

Proponents of this computer application state that the colorization process has introduced a new generation of children, who grew up with color television and movies, to classic films. Prior to this time, they may have shunned the old classics since they were accustomed to color products. Proponents also indicate that colorization does not harm the original film, and if you object to the addition of color, watch the original black-and-white version.

Opponents, in contrast, state that colorization should be halted for aesthetic and ethical reasons. While a company may have the legal right to colorize a film if it owns the copyright or the film is in the public domain, this process may be creating a distorted version of what may have been a cinematic masterpiece. Colors bleed into other colors, the lighting and makeup, which were originally designed for a black-and-white film, are now altered, and the physical details of a scene may be lost in a haze of color.

But a more salient point is the ethical issue of altering another person's work. According to many artists, including Frank Capra, who directed *It's a Wonderful Life,* no one has the right to change a piece of art, such as a film, since the artist's original vision as well as the work's integrity are destroyed. More pointedly, even if the technical quality, at some point, would make a colorized film indistinguishable from one originally shot in color, would this make colorization acceptable?[38]

Image/Digital Manipulation and the News

Another topic pertinent for our discussion is the alteration of photographs with computers. In one example, an advertiser can alter or retouch an image to enhance the picture's appearance. In another case, and one that's more relevant, a news photograph can be similarly altered.

The problem with the second scenario is the public's perception. People generally believe a news photograph portrays reality, an event as it actually occurred. By altering the image, the public's trust could be broken.

Image editing and manipulations are not, however, new phenomena. The photographic process itself, where you select a specific lens and compose a shot, is a form of editing. Image manipulations, for their part, are established darkroom fare.

Photo of Existing Conditions

Simulation of Action - Proposed Stack and Plume

New York State Electric & Gas - Clean Coal Technology Project at Milliken Station
Prepared by: Young Associates - Landscape Architecture

Figure 9–12
Computer manipulation, in this instance, is performing a valuable service, revealing the potential impact of a proposed physical plant alteration. Note the top photo of the existing conditions and the bottom photo that depicts the proposed changes. (Courtesy of David C. Young/Young Associates-Landscape Architecture)

Nevertheless, the new electronic systems, the focus of this section, raise the latter process by a notch. It's now easier to manipulate an image, and the tools required for this operation, such as a PC and graphics packages, are readily available.

In one example, you can take a picture of a site and electronically add a proposed building to the scene. In another example, you can take a picture of a swamp and electronically alter it to make it look like prime real estate.

While each manipulation produces a new image, there are differences in the context of our discussion. If the first picture is created as part of an environmental impact study, and is clearly labeled as a simulation, not many people would have a problem with this situation. In fact, it can actually provide a service. But if you don't label the swamp for what it is, that's when problems start cropping up.

Newspapers and magazines with ready access to even more sophisticated systems have been guilty, at times, of essentially turning swamps into valuable land. In two examples, *National Geographic* and the *St. Louis Post-Dispatch* respectively moved a pyramid in one scene and eliminated and replaced a can of soda in another.[39] On face value, a can of soda, in and of itself, may not be that important. Yet in the context of this photograph, it had a specific meaning.

It also raised other questions. What else can you remove, add, or *replace* in other images? Will a government, for instance, manipulate a picture to portray an event not as it occurred, but in a way that is more in line with its policies? What about abuses with legal evidence and the potential manipulation of video imagery?

This situation is further exacerbated by the emergence of electronic still photography. Instead of using film, where the original image is preserved as a negative or a slide, an image is captured and stored on a special floppy disk.

A picture can subsequently be generated in a hard-copy form with a video printer and/or can be relayed over a telephone line to another location. This capability may be particularly attractive to a news or sports photographer working in the field, since the picture information could immediately be relayed back to the home office.

This same capability, though, may make it easier to manipulate an image. It's already in a malleable form, and you don't have to bother with a print or even a negative, which in this case, doesn't even exist.[40]

Consequently, how do you prevent these abuses? One possible and partial solution may be the use of a labeling system, a symbol, to indicate that a photograph has been altered. You could also check a file's size to see if it has changed.[41] You could, that is, as long as it's the original file, and it hasn't been manipulated prior to your receipt.

Another answer may lie in the field's ethical underpinnings. Since a digital manipulation could be very hard if not impossible to detect an individual's ethical standards may be one of the only safeguards you have, and ultimately, the most important one you can trust.

Manipulation, Music, and Implications

The music industry has similarly been affected by digital manipulations, but this time through sampling. It's possible to electronically copy parts of someone else's work and to incorporate it in your own work, either with or without any changes or even the original artist's permission.

This situation is analogous to the copyright question raised in Chapter 7 with scanners. As you may recall, a scanner can copy a graphic produced by another individual. This image can then be used in a publication without the artist's permission. The same graphic can also be scanned, altered with either a graphics or image-editing program, and then used in the publication. In both instances, the artist's rights and the copyright law are violated.

Similar principles apply to the music world, and in fact, sampling has led to legal actions. Ethical questions have also been raised, and like colorization, this situation highlights one of the important attributes of the new technologies. The same family of tools that can create a work can be used to alter it illegally.

Finally, in a related topic, the proliferation of electronic musical instruments, discussed in Chapter 8, has led to its own series of problems. In one case, musicians protested, through an ad in *The New York Times*, a budgetary and/or profit-related move in which eight string players were replaced by a synthesizer. The musicians had played for a Broadway production, and the ad read, in part:

If the producers . . . really believed that a synthesizer sounds as good as real string instruments, they could have used the synthesizer to play the string parts from the beginning! But they knew that real strings sound better, so they hired eight top-notch string players to enhance the production—that's what they wanted theatre critics to hear![42]

This incident is not an isolated one, and traditional, professional musicians in various disciplines may be faced with a similar situation.

Besides the replacement of real strings by a machine, this incident touches on another topic. While a synthesizer is a machine, it is still under human control, presumably by another musician. Thus, the quality of the sound—real strings versus machine-generated music—is a factor. A similar argument about the quality of music has also been waged between proponents of CD (digital) and high-end LP (analog) record systems.

The other factor, as briefly mentioned in the ad, is the human presence of a complete orchestra. We all have certain expectations of different events. When you see a play where an orchestra is central to the production, your expectation is that it will be a conventional orchestra. It's part of the ambience of the event and experience, at least as we may currently perceive it. In a similar vein, when you go to a concert, you generally don't want to hear someone lip-synching a song.

This perception, though, could change over time. While we may object to electronic instruments in certain circumstances, the same situation may be the norm for the next generation.

Until that time, there will be a rippling effect as technological changes conflict with traditional standards. But a balance may eventually be reached. In the music world, both electronically and traditionally generated music have value. They can coexist, when used appropriately.

CONCLUSION

The broadcasting and nonbroadcasting worlds have been influenced by the introduction of computers and computer-assisted systems. They have reshaped, to a certain extent, the way we work.

Digital technology has similarly influenced these fields. The convergence of computers and video equipment has led to the development of sophisticated editing systems with which you may be able to edit digital video and then produce a conventional analog master tape.

Besides editing, digital technology has had other effects as well as implications, including

- the impact on the recording industry;
- the ability to manipulate images;
- the potential delivery of HDTV signals; and
- providing the impetus for an all-digital production facility.

The last item can be considered an evolutionary process. Digital equipment had to initially operate in what was essentially an analog sea, and an all-digital plant would reverse this situation. A videocamera's signal, for instance, could be digitized and would remain in this form until distributed. This process would streamline certain operations and could help preserve the signal's integrity.

In view of the superiority afforded by a digital plant, engineers are designing such facilities. This development also complements the creation of a fully integrated facility that can tie different equipment and systems in a centralized communications-and-control network. While stand-alone systems will continue to be used, the trend is to unite these elements, as may be the case with an automated station.

But even though these tools are powerful, they still require human input. For a graphics system, this may be the creative ability to visualize a graphic and the skill and aesthetic judgment to execute the final product.

All these developments have, as the saying goes, "pushed the envelope." Only in this case, it is a creative envelope, and one whose potential may be unlimited.

REFERENCES/NOTES

1. Robert Saltarelli, "Robotic Camera Control: A News Director's Tool," *SMPTE Journal* 100 (January 1991): 23.
2. DVE is a registered trademark of NEC America, Inc.
3. Linda Jacobson, "Mac Looks Good in Video Graphics," sidebar in "Macs Aid Corporate Video Production," *Macweek*, December 3, 1991, 40.
4. PC-generated lists have been saved on both floppy disks and on paper tape. The editing instructions could also be typed in for an on-line system. See Lon McQuillin, "The PC and Editing: One Approach," *Video Systems* 11 (February 1985): 30–34, for an interesting look at how the PC has functioned in this production environment.
5. Information from a Digital F/X demo, Spring 1992, Ithaca, NY.
6. AVID Technology, Inc., phone interview, October 1992.
7. Bob Gilbert, "Patent Impending," *Video Systems* 18 (September 1992): 4.
8. Chriss Scherer, "AKG DSE 7000," *Broadcast Engineering* 34 (August 1992): 78.
9. Brian C. Fenton, "Digital Audio Tape," *Radio-Electronics* 58 (October 1987): 78. Ironically, consumer DAT machines were actually incapable of recording a digital-to-digital copy.
10. An audiophile, like a videophile, demands the best performance from equipment—in this case, audio equipment.
11. Susan Nunziata, "DCC Technology Is Hot Topic at Winter CES," *Billboard*, January 25, 1992, 85.
12. Skip Pizzi, "Replacing the Analog Cart Machine," *Broadcast Engineering* 34 (July 1992): 46.
13. Herbert Zettl, *Television Production Handbook* (Belmont, Calif.: Wadsworth Publishing Company, 1992): 302.
14. Philip Livingston and Johann Safar, "The D-3 Composite Digital VTR Format," *SMPTE Journal* 101 (September 1992): 602.
15. The idea of security is an important one since a newsroom computer, like any other computer, may be vulnerable. The electronic newsroom should also be equipped with backup systems to maintain at least a basic level of operation in times of emergency, when the automated system may be rendered inoperable. Please see William A. Owens, "Newsroom Computers . . . Another View," sidebar in James McBride, "Newsroom Computers," *Television Engineering*, May 1990, 31, for specific information.
16. Conversation with Basys Automation Systems, October 1992. Please note: In its basic configuration, a digital still store does what it sounds like: it stores individual still images, captured pictures that are subsequently stored in memory, organized, and recalled for display.
17. The Arbitron Company, "At Arbitron, These Technologies Aren't Just a Vision, They're Reality," brochure.
18. The other systems that can fall under the ATV umbrella can range from enhanced, multichannel digital transmissions, where multiple programs are relayed on the same channel, to interactive services.
19. Tetsuo Mitsuhashi, "Scanning Specifications and Picture Quality," *NHK Technical Monograph* 32 (June 1982): 24.
20. CBS/Broadcast Group, "CBS Announces a Two Channel Compatible Broadcast System for High Definition Television," press release, September 22, 1983.
21. Elizabeth Corcoran, *Scientific American* 266 (February 1992): 96.
22. Even though there may not have been an official agreement, elements of the international production community did embrace the production standard.
23. "EIA Sets Itself Apart on HDTV," *Broadcasting*, January 16, 1989, 100.
24. Spice, also known as melange, was a rare and valuable substance.
25. U.S. Congress, Office of Technology Assessment, *The Big Picture: HDTV and High Resolution Systems*, OTA-BP-CIT-64 (Washington, D.C.: U.S. Government Printing Office, June 1990), 33.
26. Sheau-Bao Ng, "A Digital Augmentation Ap-

proach to HDTV," *SMPTE Journal* 99 (July 1990): 559.

27. "FCC to Take Simulcast Route to HDTV," *Broadcasting,* March 26, 1990, 39.

28. Frank Beacham, "Sikes: A New Ballgame for HDTV," *TV Technology* 10 (May 1992): 3.

29. "HDTV: A Game of Take and Give," *Broadcasting,* April 20, 1992, 6.

30. Frank Beacham, "HDTV at a Turning Point," *TV Technology* 9 (December 1991): 26.

31. Mario Orazio, "The Naked Truth About HDTV," *TV Technology* 10 (July 1992): 21.

32. Beacham, "HDTV at Turning Point," 25.

33. Harry C. Martin, "Rules Adopted for Implementing HDTV," *Broadcast Engineering* 34 (June 1992): 8.

34. U.S. OTA, *The Big Picture,* 83.

35. NAB, "Digital Audio Broadcasting," *Broadcast Regulation 1992; A Mid-Year Report,* 140.

36. *Ibid.,* 142.

37. Conversation with FCC, October 1992.

38. An analogy can also be drawn between this process and the controversy that erupted over the recent restoration of the Sistine Chapel's ceiling. Some critics claimed this process altered Michelangelo's original work; others stated the restoration was simply that: a cleaning and restoration process.

39. Jane Hundertmark, "When Enhancement Is Deception," *Publish* 6 (October 1991): 51.

40. *Ibid.* Please note: Electronic still cameras also have desktop-publishing applications. Special digital cameras, which store picture information in memory in a digital format, have also been manufactured.

41. Don Sutherland, "Journalism's Image Manipulation Debate: Whose Ethics Will Matter?" *Advanced Imaging* 6 (November 1991): 59.

42. Advertisement, "Grand Hotel: The Rip-Off," *The New York Times,* Section 4, November 3, 1991, 7.

ADDITIONAL READINGS

Axelrod, Marc. "Video Automation; What the Future Holds." *InView.* Spring 1990, 34–38. An overview of automation, including systems and individual units (for example, cart machines).

Broadcast Engineering 34 (July 1992) and (September 1992). Both issues are devoted to new technologies and applications in the audio and video production fields. The topics range from digital audio to DVEs. As examples, the lead articles in the respective issues are Curtis Chan, "Digital Audio Processing Production Tools," 26–30, and Ed Dwyer, "Shopping for a DTV System," 26–33.

Bruno, Susan, ed. *European Community Communication Policies: An Update.* Washington, D.C.: Center for Strategic and International Studies, 1991. A summary of highlights from a briefing with The Honorable Colette Flesch (head of the Directorate-General for Information, Communication and Culture). The report presents an overview of the EC communications policy, among other topics.

Chouthier, Ron. "Glossary of Image Processing Terminology." *Lasers and Optronics* 6 (August 1987): 60–61. As the name implies, a glossary of image-processing terms that serves as an aid to exploring image-processing techniques.

Ciarcia, Steve. "Using the ImageWise Video Digitizer: Part 2: Colorization." *Byte* 12 (August 1987): 117–21. An excellent introduction to the colorization process and how to duplicate this technique, albeit to a limited degree, with a standard PC and a special hardware accessory.

Davidoff, Frank. "The All Digital Studio." *SMPTE Journal* 89 (June 1980): 445–49. And, Nasse, D., J. L. Grimaldi, and A. Cayet. "An Experimental All-Digital Television Center." *SMPTE Journal* 95 (January 1986): 13–19. An interesting look at all-digital television facilities, from a proposal to an actual experimental digital center several years later.

Dick, Brad. "Moving into R-DAT." *Broadcast Engineering* 30 (August 1988): 63–70, 74. A look at DAT systems, including antipirating schemes.

Durrenberger, Robert S. "Microcomputer-Based Image Processing: Part 1: Fundamentals." *Laser Focus/Electro-Optics* 9 (September 1986): 126–29; and "Microcomputer-Based Image Processing: Part 2: Applications." *Laser Focus/Electro-Optics* 9 (October 1986): 130–33. Part 1 provides an overview of the microcomputer-based end of the image-processing field. Part 2 examines actual applications, including those in the medical and business fields.

Freeman, John. "A Cross-Referenced, Comprehensive Bibliography on High-Definition and Advanced Television Systems, 1971–1988." *SMPTE Journal* 101 (November 1990): 909–33. As stated, a comprehensive bibliography and a rich resource for the HDTV field.

Larish, John L. *Electronic Photography.* Blue Ridge Summit, Penn.: Tab Professional and Reference Books, 1990. A comprehensive examination of electronic photography, including still camera systems and image manipulation.

Lestarchick, John. "Editing in the Optical Domain." *Video Systems* 17 (July 1991): 14–16. A look at editing with optical media.

Pohlmann, Ken C., ed. *Advanced Digital Audio.* Carmel, Ind.: SAMS, 1991. And, Rumsey, Francis. *Digital Audio Operations.* Boston: Focal Press, 1991. Comprehensive examinations of the digital audio field, providing overviews of relevant technologies as well as practical guides.

Sloss, Teri. "Creative Creation." *Video Systems* 18 (September 1992): 20–24. A look at nonlinear editing and the editing process.

TV Technology.

"Mario Orazio" writes an ongoing column that covers a wide range of topics, including those related to HDTV.

"TV Transitions from a Global Perspective." *Broadcasting.* October 15, 1990, 50–52. HDTV and digital television from the perspective of the world community.

U.S. Congress, Office of Technology Assessment. *The Big Picture: HDTV and High Resolution Systems.* OTA-BP-CIT-64. Washington, D.C.: U.S. Government Printing Office, June 1990. A comprehensive background paper/report about domestic and international HDTV systems and related topics, such as consumer acceptance.

Watkinson, John. *The Art of Digital Video.* Boston: Focal Press, 1990. A comprehensive look at digital video. It combines an overview of the field with an in-depth discussion of more advanced topics.

Weiss, S. Merril. "Allaying the Fears Surrounding ATV." *TV Technology* 10 (July 1992): 13–14. The first in a comprehensive series of articles covering HDTV and other ATV systems.

GLOSSARY

Advanced television (ATV) systems: A generic name for higher-definition television configurations.

Audio console: The component that controls various sound elements (for example, microphones and CD players) for mixing and other audio manipulations.

Colorization: The process by which color or colorized versions of black-and-white movies are produced.

Computer-assisted editing: Using a computer to help streamline and enhance the editing process.

Digital compact cassette (DCC) and Mini Disc (MD): Two relatively new digital audio systems. As envisioned, the DCC and MD are recordable tape and optical-disc systems, respectively.

Digital audio broadcasting (DAB): A CD-quality audio signal that could be delivered to subscribers by satellite or terrestrial means.

Digital audio tape (DAT): A recordable digital tape system that can match a CD's audio quality. DATs have also been used as storage devices for computer data backups.

Digital effects generator: A production component that digitizes and manipulates images to produce visual special effects.

Electronic newsroom: A newsroom equipped with computers for newswriting, creating databases for news morgues, and directly feeding copy to prompters, among other operations.

Graphics generator: The component that is used to create computer graphics. The graphics can be produced either with special, dedicated graphics stations or properly configured PCs.

High-definition television (HDTV): The field concerned with the development of a new and improved television standard. Superior pictures will appear on a wider and larger screen; both digital and analog configurations have been developed.

Image processing: Image processing is the field and technique in which an image, typically created by a videocamera, is digitized and manipulated. Typical operations include image enhancement and correction.

Microprocessor-assisted equipment: A video or audio component with a built-in microprocessor that can automate various tasks.

Nonlinear editing: You can gain access to video segments in a nonlinear fashion, speeding up the editing process.

Photograph and image manipulation: The manipulation of photographs for, in one example, electronic retouching (used in advertising). A potential problem is the alteration of news images.

Switcher: The video component that is used to select the various camera and videotape sources in the course of a production. It can also be used in post-production work and may incorporate computer technology to assist in manipulating the different video sources.

10 Information Services: Teletext and Interactive Systems

This chapter examines teletext magazines and two-way interactive systems. While the focus is on dedicated information services, which could potentially be offered as a video dial tone option, various computer-based communications systems are also explored.

The chapter concludes with an overview of the entire spectrum of computer-based services with respect to their legal and political implications. The issues range from privacy to First Amendment rights to the democratization of information. The latter topic is complementary to the discussion in Chapter 7.

TELETEXT ELECTRONIC MAGAZINES

A *teletext magazine* is an electronic publication delivered to your home television set either through a television broadcast signal or a local cable system. It is composed of text as well as graphics and is read as it appears on a television screen, similar to the way a printed magazine is read.

Like its paper counterpart, a teletext magazine is composed of news, sports, and special interest features. There are generally 100 or more pages of information, and a topic index organizes the stories into specific subject areas.

You use a standard television set, a decoder, and a keypad to actually gain access to the magazine. The keypad resembles a television remote control unit and is used to select the page you want to view.

A teletext magazine can also be classified as a *one-way information stream* that flows from the originating company to its subscribers. This is in contrast with the two-way interactive systems examined later in the chapter.

Teletext Magazines and Their Services

Teletext magazines have several unique capabilities. In one example, a teletext magazine could complement closed-captioned programming since its stories appear on the screen as written text. These stories could also be regularly updated on a 24-hour service, unlike newspapers and most commercial broadcasting outlets that have specific daily deadlines.

The business and educational communities could also be supported by a teletext magazine. A service could feature local ads, and for education, it could be an active force in the school system. The latter was exemplified by the work conducted by KCET-TV in

Figure 10–1
An index page of a teletext magazine. (Reprinted by permission of ELECTRA, a part of Great American Broadcasting; Electra)

Los Angeles, California. In the words of Dr. Ronald J. Goldman, the director of teletext services when the project was operational, "we believe teletext possesses enormous potential to change the very nature of the television viewing, from a passive to an active experience, . . . to be a source of substantive and constantly-available information; and to make television a truly interactive tool."[1]

KCET-TV developed a range of programming, including an interactive component that addressed the needs of an elementary-school curriculum. By using a teletext system, the station demonstrated that a television set was no longer just a box to watch passively. It became a teaching tool.

Teletext in Europe

In 1972, engineers working for the British Broadcasting Corporation (BBC) developed and tested *Ceefax*, the world's first teletext magazine. By the mid- to late 1970s, it developed beyond the experimental stage into a fully operational system. About the same time, a second teletext magazine, *Oracle*, was created by the Independent Broadcasting Authority (IBA) in England. The two competitors, along with British television manufacturers, also established a common teletext standard for England.

As the British refined their operations, France produced *Antiope*, its own version of a teletext magazine, while Canada introduced the *Telidon* standard. Telidon was designed to accommodate both teletext and interactive two-way services. Other nations, including the United States, similarly experimented with and implemented their own systems.

Teletext in the United States

Teletext operations in the United States have not been as widespread as those in other countries. One reason for this condition is the economic structure of the American television industry.

Unlike England, for example, with its strong noncommercial television system, the United States is dominated by commercial television interests. Consequently, the technology did not gain widespread support, since the networks originally believed a teletext magazine would stop viewers from watching commercials.[2]

But experimental services were eventually launched, and after a developmental period, two teletext standards emerged in the United States. The first, *World System Teletext* (WST), was based upon British operations. U.S. advocates of this system have included Great American Communications, the producer of the *Electra* teletext magazine, and Satellite Syndicated Systems, a cofounder of *Keyfax*, a national teletext service. The second standard, the *North American Broadcast Teletext Standard* (NABTS), was developed through the combined efforts of CBS, the Canadians, and the French. Domestic supporters of this standard have included CBS (*EXTRAVISION* magazine) and NBC.

Two teletext standards were in force in the United States partly due to a 1983 FCC ruling. The commission selected neither WST nor NABTS technology and declared that market force competition would determine the eventual winner. Basically, if there was enough support for two distinct teletext standards, two standards would continue to exist.

This regulatory philosophy had positive and negative results. On the positive side, the FCC gave the media organizations that produced teletext magazines, as well as consumers, the option to judge and choose the most appropriate standard.

On a negative note, the presence of two incompatible standards splintered the potential teletext market in the United States. Since a single standard was not adopted, manufacturers had to support one of the two standards in the hope that it would eventually dominate the industry. In reality, though, few were willing to take this financial and marketing gamble.

The absence of a single standard also prevented the teletext industry from creating a strong foundation in the traditional communications system. It lacked focus, and the potential market splintered, which in turn blunted the industry's ability to become an established media presence.

This situation was different from that of stereo television, another area where the FCC did not officially mandate the adoption of a particular format. But a de facto standard emerged, which helped accelerate the integration of this technology.

Teletext Distribution

A teletext magazine can be distributed via a television signal's *vertical blanking interval* (VBI). Videocameras and television sets produce and reproduce their pictures through complementary and synchronized scanning cycles. During this process, a camera tube's light-sensitive element, the target, is scanned by an electron gun in a precise pattern and in a series of 525 steps or lines.

The electron beam initially scans the odd-numbered lines, reaches the bottom of the tube, and returns to the top to scan the even-numbered lines. When the beam travels to the top of the tube, a period of time elapses in which the beam is blanked or turned off. This is known as the VBI.

Teletext services have taken advantage of this television-scanning characteristic by inserting their electronic information in the VBI.[3] Consequently, teletext magazines are piggybacked or multiplexed with standard television signals and do not require separate broadcast signals for their transmission.

Next, the television signal is relayed, either by satellite or terrestrial means, and is eventually received by the teletext decoder. A subscriber can then use a keypad to retrieve and view the magazine: a page can be selected, and the data that compose the page are removed from a data stream by the decoder for eventual display on the television screen.[4]

A page could include text and graphics, the latter defined by either an *alphamosaic* or *alphageometric* system. WST magazines, such as *Keyfax*, supported alphamosaic graphics, which were more blocklike in appearance than their NABTS-based alphageometric counterpart. However, an NABTS decoder was more expensive than a comparable WST

unit, in reflection of this enhanced graphics capability.

Future Development of Teletext Services

Besides the services outlined in this chapter, such as the delivery of international and national news stories, teletext magazines could support additional features. Time, Inc., for example, had inaugurated a 5,000-page teletext or, in this case, a cabletext magazine, that would have been nationally distributed via satellite to cable companies that devoted one channel to this venture. The magazine's large page capacity would have supported in-depth stories and other features. Even though this operation never fully materialized, it demonstrated that a teletext service could distribute thousands of pages of information in contrast to the typical 100-page operation.

Another proposal, briefly mentioned in Chapter 8, would be the wide support for the transmission of specific news and information to only select subscribers. In this system, a subscriber could define and receive an individualized information pool, special-interest features, besides the magazine's standard stories. This selective transmission would be made possible through addressable decoder boxes, analogous to the ones described in Chapter 5.

This scenario would reflect the trend in print magazines. If you look at any newsstand or magazine rack, you will see specialized magazines about computers, cars, photography, and other topics. Teletext magazines could possibly duplicate this more personal rather than mass communications service with addressable boxes.

New decoders could also support higher-resolution images and an enhanced user interface to speed up the access to various stories.[5] In a related area, more television sets with internal teletext modules could be built. An internal decoder could make the technology more transparent to the viewer, since a magazine would always be available without the bother of buying and connecting an external device. This ease of use, when combined with a very low distribution fee—or no fee at all—would give a consumer an added incentive to become an active teletext user.

The industry could receive a further boost through narrowcasting and slanting programming toward hospitals and other business and professional interest groups. A teletext magazine also remains an attractive distribution medium, in view of its spectrum efficiency, and a service could still be relatively cost-effective to implement.[6]

Conclusion

Despite this past and future promise, teletext magazines have not emerged in the United States as a viable communications tool. Part of the problem was the presence of two competing standards, the cost for decoders, and possibly the high level of media saturation in the United States. Unlike another country where a teletext magazine may have been government supported and media sources may have been limited, the teletext industry in the United States was a commercial enterprise. It had to compete with a wide array of entertainment and information programming, including specialized television and cable shows and the computer-based information systems described in the next section of this chapter.

Perhaps the average consumer did not believe a teletext magazine was an important and desirable programming option. In response, the industry may have had to more fully demonstrate the value of its product and/or focus its effort on select user groups.

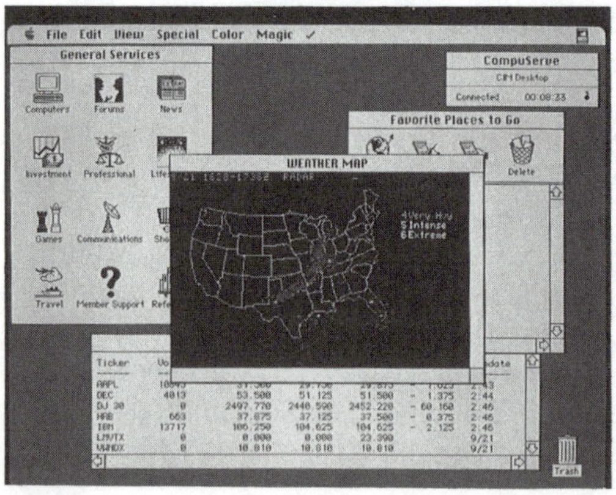

Figure 10–2
A PC-based system may support an enhanced user interface to speed up and simplify an online session. (Courtesy of CompuServe, Inc.; CompuServe Information Manager)

Finally, in this book's first edition, it was indicated that the teletext industry was at a crossroads. It had the potential to develop into an important element of the domestic communications system if a number of events took place. These included the widespread use of television sets with built-in decoders and a solution to the standards problem. As of this writing, though, it appears the crossroads may, in reality, have already been passed by.

TWO-WAY INTERACTIVE SYSTEMS

This second half of the chapter focuses on interactive two-way information services. Other computer-based configurations are also examined, as are various legal and political issues.

Introduction

The key element in any two-way operation is the capability to interact with a system. Unlike a teletext magazine's one-way information relay, where the same 100 or more pages are delivered to all subscribers, a two-way interactive system is designed to meet an individual's needs. In essence, the subscriber requests and receives specific information.

Two-way interactive services can also be divided into three classifications: *videotex*, sometimes spelled *videotext*, operations; *PC-based interactive* operations; and *dedicated information-retrieval* operations. The classifications are differentiated by the types of information and services a given system may support, the equipment that forms the communications link, and the targeted subscriber group. In the context of our discussion, this three-tiered scheme simply provides a framework to explore the interactive universe.

Videotex Operations

A videotex system is a two-way, graphics-oriented, interactive service that supports a database comprising thousands of separate frames or pages of information. A subscriber can retrieve this graphic and alphanumeric information from a computer via a standard telephone line, the prevalent communications channel for this type of operation and, in fact, for the other systems described in the chapter.

Different terminals have been developed to create this communications link. A typical configuration consists of a decoder interfaced with a television set, the viewing device in this setup, and a keypad, analogous to the keypad employed in a teletext system. A stand-alone terminal, complete with a screen, keyboard, and an internal decoder, has likewise been available.

Besides storing a wide range of information, a videotex system can support various information-based services and interactive *transactions*. These may include banking and shopping at home.

A videotex system is distinguished by an easy-to-use interface and one that simplifies the connection process for the subscriber. For example, the modem is already provided as part of the configuration, unlike a PC-based system, in which a user may have to purchase and set up this device as well as the prerequisite communications software.

The videotex systems in this chapter have also been generally designed to satisfy perceived consumer needs. In light of this slant, a system may have included movie reviews, an electronic encyclopedia, and shopping at home. Colorful graphics have also been used to convey information and to attract, inform, and interest a subscriber.

But videotex systems have likewise supported the business community with an assortment of special functions, as we shall see when we examine the *Prestel* and *Viewtron* systems.

Prestel. The world's first public videotex system, Prestel, was launched in England in 1979. This comprehensive system has been distributed throughout the country.

Unlike the free teletext information, a subscriber is assessed a fee to use the service. The data are stored as frames or pages, much like the pages in a book or in a teletext system.

An *information provider* (IP) could also charge a subscriber for retrieving specific information. An IP is an organization that contributes information to the database. An airline IP, for example, could provide specific flight and price information as well as the available standby seating for that day's flights. Other IPs have supplied financial data, newspaper stories, and an assortment of information.

Prestel has incorporated a series of indexes, menus, to assist a subscriber in information search and retrieval. The menus can guide a subscriber from broad to narrow subject areas; this helped make Prestel a somewhat user-friendly system, since it gradually led an individual to the desired information. A subscriber familiar with Prestel can also immediately view a given page by punching in the appropriate number with a keypad or a keyboard.

Besides serving the consumer, Prestel has supported the business community. In one application, an IP could set up tailor-made data pools for different departments in a large organization. Access to the information could then be restricted to all parties other than designated individuals.

Another important feature supported by Prestel has been a *gateway*. A gateway enables an individual to gain access to different and normally incompatible and inaccessible computer systems outside of the host system. It is, to all intents, a gate, a link, to these other operations.

Travel agents have, for instance, used this option to interact with airline computers to obtain travel information and to make reservations.[7] In a related scenario, a similar service called Homelink was implemented to serve the general public. Individual subscribers could, as an option, conduct electronic banking transactions.

In sum, Prestel has provided England with a comprehensive and sophisticated interactive information network. On one level, subscribers have been able to locate specific data through either a series of indexes or by simmediately punching in a page number. This same system could also link Prestel users with different computers and databases of information via a gateway option, and Prestel has likewise functioned as a communications and information center for the business community.

Viewtron. In contrast with England's national Prestel service, a number of separate videotex operations have been introduced in the United States. One of them was Viewtron, a consumer-oriented venture that was inaugurated in 1983 in Florida by the Viewdata Corporation of America.

Viewtron was a comprehensive service that offered weather and marine forecasts, games, and a hookup to the local library. The system also carried a wide range of IPs, including *The New York Times* and an electronic television guide, *TV Data*.

One of Viewtron's most appealing features was a gateway option used for electronic banking and other functions. A subscriber could review and purchase merchandise from the J.C. Penney Company, among various stores, and could retrieve information from *Grolier's Encyclopedia*.

If a subscriber ordered a product from a local business hooked into the system, the merchant could send, in return, an electronic receipt for the order. This receipt was then stored in the appropriate Viewtron mailbox, a data-storage area reserved for each subscriber.[8]

This ability to conduct an interactive transaction with a local business highlighted one of Viewtron's strengths. It was designed to be fully integrated within a given community. Local businesses and even public institutions, such as libraries, could join the network to establish an interactive community-wide information system. This local network was further enhanced by the Viewtron pages exclusively devoted to local news items.

A subscriber could gain access to Viewtron with Sceptre, a terminal connected to a telephone line and a standard television set. Indexes were supplied by Viewtron to guide the subscriber through the pages of information, and the system was configured to accept keywords such as *weather* and *sports*, so the subscriber could quickly locate specific data.

Despite these features, consumers did not support Viewtron. Part of the problem was the terminal's initial high cost and the service's user fee. Consequently, to make Viewtron available to a broader user base, the service was eventually opened up to PC owners across the country who were equipped with the proper software. By taking this step, and others, Viewtron was essentially transformed from a community-oriented service to one that had to compete with *CompuServe* and other established PC-based companies.

Finally, the service was terminated in early 1986, after a financial loss.[9] About the same time, another consumer-oriented videotex project, Times Mirror's *Gateway*, was similarly canceled.

The cessation of both systems, which were supported by some of the larger companies in the communications field, seemingly highlighted a developing trend in the United States: the average consumer did not view videotex, and possibly in a broader sense, computer information services, as a priority item. Although Viewtron and similar systems were not targeted toward all consumers, due to their features and price structure, they nevertheless failed to attract enough consumers who could afford or wanted to become subscribers.

Telidon. In the context of this chapter, Telidon can be viewed as both a videotex system and standard. Unlike Prestel, Telidon has accommodated a number of separate videotex services that conformed to the standard devised for the overall Telidon system. As such, all of these systems and, in a sense, their complementary teletext configuration, can be construed as members of the broader Telidon family.

The Telidon project was the brainchild of the Canadian Department of Communications. When first envisioned, the designers' goal was to create, in part, a videotex system free from hardware constraints. Older terminals would be capable of accessing "future command formats," while newer terminals would be compatible with old or preexisting data.[10] If a terminal received a command to draw a line on the screen, for example, a newer terminal could draw a high-resolution line; an older terminal that lacked this feature would still be able to interpret and execute the command, albeit at a lower resolution.

One of Telidon's most powerful features has been its support of an alphageometric coding scheme, and thus, alphageometric graphics. In this configuration, *picture-description instructions* (PDIs) essentially define various geometric shapes as they would appear on the screen. A Line command, for one, would specify two endpoints on a screen, subsequently connected by the terminal to form the line.[11] This approach simplified the graphics creation process, sped up the information relay, and helped ensure that an operation would be compatible with new developments for a number of years. In this respect, the system is analogous to the Post-Script language adopted by the desktop-publishing industry.

Telidon's valuable features, including the compatibility between older and newer system designs, led to its role as a model for the videotex standard announced by AT&T in 1981. Following this decision, the Canadian Standards Association and the American National Standards Institute jointly developed the *North American Presentation Level Protocol Syntax* (NAPLPS), the standard based upon the AT&T and Telidon concepts. It was subsequently selected for the Viewtron system, among other operations, and embodied PDIs and additional refinements.

As indicated, the Telidon standard served as host for a variety of videotex projects throughout Canada. This included Grass-roots, a system in Manitoba that provided farmers with commodity price, local weather, and community information. Another videotex configuration, Teleguide, carried free tourist information on a series of public terminals.[12]

PC-Based Operations

The services provided by Viewtron and others have generally been designed for individuals who may not have been computer-oriented, even though some subscribers were computer hobbyists. In view of this philosophy, the hookup procedure was simplified, the videotex system may have been integrated within a given community, and graphics were widely used.

Companies such as CompuServe, on the other hand, are more nationally oriented in their subscriber base, and geared toward subscribers who are either familiar with or owners of PCs. This also holds true for *Prodigy*, a service that could be categorized as a cross between a more traditional videotex operation and a PC-based system.[13]

While PC-based configurations provide some of the same services as their videotex counterparts, the information is usually text-based, and almost any PC owner can gain

Figure 10–3
The typical connection between a home user and an information company is the telephone line. In general, a telephone line and modem are the "bread-and-butter" connection of current interactive services.

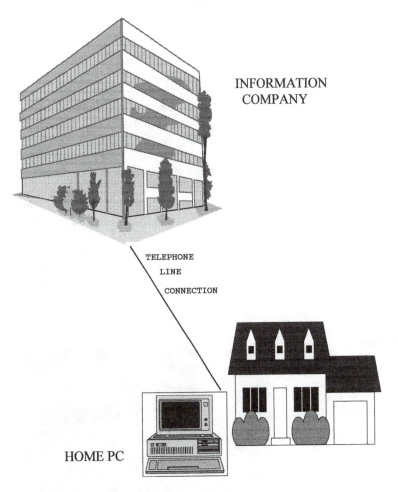

MODEM CONNECTION

INFORMATION COMPANY

TELEPHONE LINE CONNECTION

HOME PC

access to such a system with an inexpensive modem and communications software. This has not been the case with the videotex industry, where multiple standards have prevailed.

The information is actually arranged in databases designed to be retrieved with a computer's keyboard. A subscriber can type in a specific request for information and may have to contend with only the occasional menu, even though a series of menus may be available for individuals who are not familiar with the system. A PC-based system may also support an enhanced interface that simplifies navigation through the various information databases and services.

Finally, while a PC-based system may rely on text for much of its communication, it can also support graphics. An operation such as CompuServe is, for example, a graphics-rich warehouse of images, animations, and other visual information. Different graphics formats, for different PC platforms, are also supported.

CompuServe.   CompuServe has been one of the largest PC-based systems in the United States. As such, it is used to illustrate an operation of this nature.

You are actually connected to CompuServe through a PC, modem, and the telephone line. Initially, a user was charged a variable fee for the *connect time*, the number of minutes or hours the subscriber was online with the service. This fee was affected by the modem speed and other factors.

But following a pattern popularized by Prodigy, a flat-fee option was later adopted, where you would be charged a specific monthly rate regardless of the actual connect time. In CompuServe's case, though, this option initially covered a select range of offerings. For those options not included in the flat-fee rate, the standard rate card applied.[14]

As of this writing, CompuServe features a wide array of on-line databases as part of the standard service. On-line implies, in this situation, information you receive while you are linked to and interacting with the company's computer. A sample of databases includes national and international news, income-tax information, travel guides, and topical databases for events such as the Olympics.

Besides the standard fee, an additional charge may be assessed for retrieving other, more specialized data pools. Stock price quotes as well as demographic information, statistical data that profiles the attributes and characteristics of a given community, fall under this category.

A number of *forums*, also known as special interest groups (SIGs), are featured on PC-based systems. A forum, which may be considered an electronic meeting place, may be dedicated to any one of a number of PCs, such as IBM or Apple computers, in reflection of the nature of the PC-oriented subscriber base. A forum usually supports a special files area, where electronic messages may be either written or read, and a software section(s).

The latter can include entire libraries with programs that can be downloaded and retrieved by a subscriber. Some of the options include public-domain programs, software utilities, games, and computer graphics and animation collections.

A forum may also sponsor a computer conference, where subscribers can communicate with each other, through the system, on pertinent topics. In one particular application, experts in a programming language, such as Lisp, may answer a series of electronically submitted questions or may present information about the language.

Forums are also dedicated to topics that are not generally related to the computer field. They may be concerned with specific hobbies, health issues, and for the communications field, broadcast engineering and journalism.

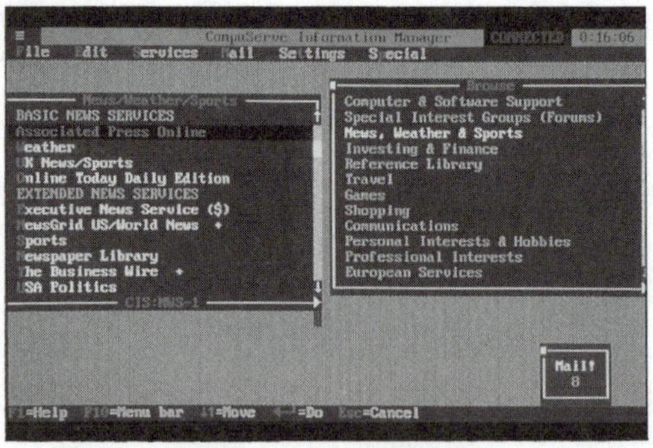

Figure 10–4
Some of the information products offered by PC-based systems. (Courtesy of CompuServe, Inc.; CompuServe)

In addition to forums and other information-related services, CompuServe features an electronic mall where you can buy everything from coffee to a computer. CompuServe also supports E-mail, both internal and external to the system, and gateways. You can find a book in a bibliographic search, look at different airline flight schedules, and conduct a sophisticated information search to locate and, if so desired, retrieve a series of articles in support of a research project.

Industry is likewise supported by CompuServe and PC-based systems. A company can subscribe to an enhanced service where a series of business-related databases are featured. There is also a strong interest in electronic business and financial transactions, such as buying and selling stocks. In fact, *Dow Jones News/Retrieval*, another and more specialized PC-based system, has been particularly strong in this area and has offered many business- and financial-related options.

Dedicated Information-Retrieval Operations

The last category of two-way interactive operations, the dedicated information-retrieval system, consists of companies that support small and highly specialized subscriber groups. A dedicated system is typically accessed with a PC and offers subscribers unique pools of information, including those applicable to the legal and medical professions.

As the name implies, a dedicated information-retrieval company is strictly in the information business. It is differentiated from the PC-based service, for our purposes, by the targeted user groups and supported information databases.

Lexis and Nexis. Two of the more interesting members of this classification have been Mead Data Central's *Lexis* and *Nexis* services. Lexis is a computerized legal database that provides legal citations and print-outs of court cases, among its functions. It is also a valuable information resource, considering the growing importance attached to the international community. Lexis has covered specific elements of other countries' legal systems in addition to our domestic system.

Nexis, on the other hand, is a news-retrieval service that has been adopted by television station news departments and other organizations. It is a journalistic tool with an information pool composed of more than 100 newspapers, magazines, and specialized publications and transcripts.

Even though the Nexis service can be expensive, it can save valuable time, as is the case with Lexis. This can be a critical factor, more often than not, in the news business. For example, rather than spending half a day conducting research in either a news morgue or a library in preparation for a detailed report about the Environmental Protection Agency, a journalist can locate and retrieve full-length articles about the subject in a matter of minutes.

It has also been possible to use Nexis to obtain an international slant about a topic, by retrieving information from international markets. This option may prove to be valuable when writing, in one case, a news report about a series of economic summits held between the United States and Japan. Rather than solely concentrating on the American viewpoint of the meetings, it would be possible to obtain an international perspective. This information may prove to be interesting and may shed additional light on not only the meetings, but also the international community's reactions to any trade agreements that were signed at the time.

In addition to Mead Data Central, other companies or information vendors, including BRS Information Technologies, Dialog Information Services, Inc., and NewsNet, Inc., offer specialized information databases. These may range from entire collections of scientific abstracts to detailed records that can highlight a company's financial history.

NewsNet is also particularly appropriate for the communications field, in view of the breadth of the information resources that are made available through its system. Topics have included satellite and fiber-optic communications, the FCC, and optical discs.

NewsNet has also supported, as have some other services, an *electronic clipping* function. Rather than tediously searching through a database to spot specific news and information items, the system could be set up to complete this task. A subject area is initially defined. Next, the relevant news from a database, which is constantly updated, is collected, stored in an electronic mailbox, and eventually retrieved by the subscriber.

OTHER CONFIGURATIONS

Besides NewsNet and the operations just described, another type of computer-based communications system, a *bulletin board system* (BBS), can be created. A BBS typically consists of a PC equipped with a hard drive and special software that can transform this configuration into a miniature two-way interactive system. A BBS can support a database of information and services, and its capabilities can vary widely.

Typical options include extensive public-domain and shareware software offerings.[15] Messages can be delivered to public and private file areas, and BBSs have been used to distribute electronic publications.

BBSs have also been set up by manufacturers to serve as information resources. Product questions can be quickly answered, and software updates can be made available to customers. Other BBSs have been created by government agencies, such as NASA, to distribute information.

Computer owners who are simply interested in this type of venture have also set up BBSs. This type of operation, which may be free or carry a user fee, is an example of a grass-roots computer and information system. In fact, since a BBS is a fairly cost-effective proposition, literally hundreds of personal BBSs have sprouted up across the country.

Some BBSs are also very comprehensive and can equal, in certain respects, the systems described in the previous two sections. *Channel 1* in Massachusetts, for example, which could be more aptly called a conferencing system, has supported multiple users, over 60,000 archives, on-line multiple-user games, and a link to Internet. This is in addition to its general services and information resources.[16]

As with other tools of the communication revolution, some BBS owners have also been involved in legal issues. In one case, some BBSs have illegally carried pirated software. Other issues touch on raising user fees for telephone-line access and the distribution of allegedly valuable proprietary information. These topics are discussed in the last section of this chapter.

Finally, besides BBSs, other computer-based configurations as well as networks will continue to play a growing role in our communications system. These include the Internet, as briefly described in Chapter 3, operations such as the *WELL* in California, community-based systems, and *Bitnet*.[17]

Bitnet is an international "academic" network that has supported E-mail, file exchanges, and other services. It is also a host to discussion lists, essentially groups of people interested in the same topic. This includes space program enthusiasts, who are served by Space Digest.

At its core, Space Digest functions as a communications link, an information clearinghouse, and an electronic forum. Its topics vary widely and, on a typical day, can include discussions about private space ventures, commercial space developments, and NASA's proposed space station.[18]

More in line with the communications field, Bitnet has provided users with an access route to *Comserve,* a forum that supports information databases geared for communications professionals, educators, and students. It has also featured, among its options, a discussion group that focuses on mass communications and the new technologies.

Summary

When viewed in its entirety, the computer-based communications universe is an important resource. You can send E-mail to a friend or colleague, search for a book in a library located in another state, and communicate with another person, via the keyboard, either in real time or not.[19] You can also join special-interest groups, conduct research for a project, buy and sell stock, retrieve electronic journals, and use the Internet as an electronic highway to tie into an ever-growing collection of resources.

The systems that make up this universe, which include two-way interactive and teletext operations, have also contributed to the growth of our paperless society. As discussed in Chapter 4, a paperless society is one in which information is increasingly created, exchanged, and stored in an electronic form.

Yet while many of these paperless or electronic resources are free, others are relatively expensive. Like a thirsty swimmer surrounded by salt water, a person seeking a bit of data from the wealth of information available may find it financially unreachable. The knowledge and skills that may be required to tap this information only add to the problem.

This scenario brings up the same question that was asked in previous chapters. If information is reduced to an electronic form, will there be wide access to it? If not, what are the potential implications?

A partial solution may be offered by examining *Teletel*, also known as *Minitel*, the French interactive information system. In an attempt to promote this national service and to lower printing costs, the telephone directory was installed on the network, and the government distributed free terminals to French citizens. This national system has subsequently flourished, and while it is not without its own set of problems, it has supported a wide range of information services. Another solution may take the form of a nationwide network, as described in the next section of this chapter.

LEGAL, SOCIAL, AND POLITICAL IMPLICATIONS OF COMPUTER-BASED COMMUNICATIONS SYSTEMS

To wrap up this chapter, it's appropriate to examine some of the implications raised by computer-based communications systems. These include privacy questions and the democratization of information.

Growth of CompuServe

(chart: values 200K, 400K, 600K, 800K, 1000K; years 1979 1980 1981 1982 1983 1984 1985 1986 1987 1988 1989 1990 1991)

Privacy

E-mail.

Thinking of electronic information differently from, say, a personal letter at the post office is natural. The intangible character of electronic information makes the attributes of privacy and ownership somehow seem less real. So we often take for granted how, for example, we might react to someone reading private E-mail. Yet doing so is no different from opening a sealed envelope that's been stolen from the post office.[20]

The privacy question is of paramount concern in our new communications universe. As outlined in Chapter 4, CD-ROMs packed

Figure 10–5
This diagram shows the growth of CompuServe (number of members) from 1979 to 1991. The increase, from 1,200 members (when the CompuServe Information Service began) to 900,000 members in 1991, also highlights the growing popularity of this type of PC-based service. (Membership data courtesy of CompuServe, Inc.)

TRADITIONAL MEDIA

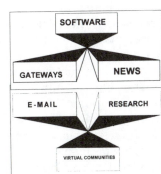

SAMPLE OF THE SERVICES OFFERED BY COMPUTER BASED INFORMATION SYSTEMS

Figure 10–6
Sample of the services offered by computer-based information systems. There is also a convergence between these systems and the traditional media (top). In one case, as described in Chapter 5, cable companies may support information services in addition to entertainment programming.

with consumer information and Caller ID services have the potential to be invasive, since our privacy could be violated. Another important area is that of E-mail.

The topic of E-mail was first introduced in Chapter 3. Electronic messages can be relayed on a LAN within an organization and to outside locations through public conduits provided by networks such as CompuServe.

Concerns have been raised about the privacy of this information. As the volume of E-mail increases, so does the potential for abuse. Just as someone can read a conventional letter, the security of an electronic mailbox and the information inside could be breached, under certain circumstances.

The 1986 *Electronic Communications Privacy Act* sought to address this issue. It didn't, however, cover E-mail within a company, and like other forms of communication, an employee's E-mail could potentially be read.

This situation raises important questions. A company may indicate that employee communication should be open to monitoring to ensure the company's resources are not being used for personal gain. But if monitoring does take place, are employees informed that their E-mail, in this particular example, may be read? If they aren't notified, and privacy is assumed, what are the legal ramifications if monitoring takes place? In one case, this factor played a role in a lawsuit.[21]

Like the colorization issue discussed in Chapter 9, there is also an ethical side to this question. Even if a company has the legal right to read an employee's E-mail, is it ethical to do so?

Wiretapping, Privacy, and Digital Technology. Another area that falls under our discussion, and is included in this chapter for organizational purposes, is electronic wiretapping. In the past, the structure of the telephone system made it easier to tap someone's line. But the introduction of digital communication and FO systems, where a voice may now be digitized and relayed in the form of light, makes electronic eavesdropping more complicated.

The situation is only exacerbated by the capability to encrypt information in a code so it is unintelligible without the proper decoding mechanism. This is analogous to an encryption scheme used to protect a satellite's relay.

For most people, the added security is welcome news. The FBI and other law-enforcement agencies have, however, taken the opposite stance, since their job has become more difficult. Consequently, legislation has been proposed to ensure that these organizations have continued access to the information. In one proposal, an electronic back door, which would allow an agency to gain access to the targeted information, would have to be built into a communications system.[22]

Elements of the business community and other organizations have been opposed to this type of legislation on two grounds, privacy and economics. While a court order must be obtained to initiate a wiretapping procedure, as has been the practice, legislation may make the communications network more susceptible to illegal tapping. The system would also be more open to potential governmental and nongovernmental abuses. Ultimately, instead of using technology to build a more secure communications system, which could carry voice as well as computer transactions, the system could potentially be compromised.

A network could be more expensive to build, and a manufacturer would have to contend with more red tape. Other concerns have centered on the stifling of technological developments and the international implications of such legislation.[23] As other nations developed more secure networks to carry the world's communications traffic, the U.S. systems could be burdened by this organizational hierarchy and a communications structure that could be vulnerable to government and potentially illegal eavesdropping.

The government, for its part, has defended its stance in the name of law enforcement and national security. The prosecution and conviction of a criminal could, for example, hinge on a recorded telephone call(s). The inability to tap this and other forms of information would hinder such an operation. It was also argued that this authorization was an extension of a 1968 act that permitted "law enforcement to have court-ordered and court-authorized access to private-normally private information . . . if they involve criminal conduct."[24]

Regardless of the side you may support, this debate does highlight another fallout of the communication revolution. Technological changes and the resulting structural impact may take place so rapidly that the legal system may not be able to keep pace. In this instance, a concern over the potential difficulty or inability to tap into the new digital and optical data stream, and to isolate specific information such as a telephone call, was the basis for the proposed legislation. It is also important to note, as discussed on ABC's "Nightline," that the Department of Justice could play the major role in defining a new standard, not the FCC.[25]

This raises two questions. Are there additional areas in the *commercial* communications arena where standards and operational parameters may be defined by bodies such as the Department of Justice? If so, what are the ramifications of this development?

Other Issues

Besides the privacy questions, other issues are relevant to our present discussion.

Communication Fees. In 1983 a regional telephone company tried to institute a modem connect charge. Customers who linked a computer and modem to the telephone line would have been assessed an additional monthly fee.[26]

Even though this surcharge was later dropped, the implications remain. If a fee, especially if it was high enough, were ever permanently established in this country, it could wreak havoc with two-way interactive companies and other organizations that use the telephone system for this type of communications venture.

Similarly, the FCC geared up its regulatory machinery in 1987. The goal was to institute a rule change that would require service providers, such as CompuServe and the other players in this field, to pay an access charge for their links with local telephone networks. This fee was originally waived to accommodate the newly developing infor-mation industry that was dependent upon the telephone system for its survival. But the FCC believed the industry had matured to the point where, by the late 1980s, the exemption was no longer warranted, and it was unfair for the local telephone companies to continue to absorb the additional expense.

This proposal triggered a strong reaction on the part of the public and the companies that would be affected by this plan, and in 1988, it was dropped under the threat of congressional intervention. Nevertheless, the potential ramifications raised by this affair are similar to those of the modem scenario. If a fee schedule is introduced at some future date, it could add a monetary surcharge above and beyond the standard rates. This factor could make a two-way interactive service too expensive for many individuals.[27]

Content Issues: Prodigy, *Phrack*, and CompuServe. Besides potential communications fees, computer-based system operators have contended with content issues. Prodigy, for one, was scrutinized for possible First Amendment violations for editing or deleting controversial messages from its bulletin boards.[28] The company responded by indicating it retained editorial control over its electronic publication, much like a conventional newspaper.

In a related area, Craig Neidorf, a publisher of the electronic magazine *Phrack*, was arrested and his operation was raided by the government for publishing a telephone document that purportedly contained sensitive material. The information was actually obtained from a telephone company's computer by another individual, and Neidorf was "indicted on felony charges of wire fraud and interstate transportation of stolen property."[29] But when it was revealed that the same information was publicly available from the telephone company for a small fee, the prosecution dropped the case.

Neidorf's and related cases, though, brought the opinions of two opposing camps into focus. On one side, law-enforcement agencies have upheld their duty to prosecute anyone who illegally gained entry into a computer system, which in Neidorf's case ultimately resulted in the electronic publication of a stolen document. On the flip side, defenders of both Neidorf and individuals involved in similar cases indicated that publishers and their electronic publications should be afforded the same protection as the traditional print media. Furthermore, while illegal computer entry should be prosecuted, the punishment should fit the crime and the individual's intent. It shouldn't be disproportionate to the true extent of the damage an illegal intrusion may have caused.[30]

CompuServe, for its part, has also been involved in various content issues. In one case, the company was sued over allegedly defamatory statements made on one of its forums.

But in U.S. District Court (Southern District of New York), the judge found that CompuServe was not liable for this information. It could not be shown that CompuServe knew or had reason to know about the alleged defamation. In rendering his decision, Judge Leisure held that

While CompuServe may decline to carry a given publication altogether, in reality, once it does decide to carry a publication, it will have little or no editorial control over that publication's contents. This is especially so when CompuServe carries the publication as part of a forum that is managed by a company unrelated to CompuServe. . . . CompuServe has no more editorial control over such a publication than does a public library, book store, or newsstand, and it would be no more feasible for CompuServe to examine every publication it carries for potentially defamatory statements than it would be for any other distributor to do so.[31]

This outcome was viewed as a victory by free speech proponents. They feared an adverse ruling could have had a chilling effect on the entire industry.[32] In this particular instance, if a company was liable, would it have to police all the information it carried? This could be an almost impossible task and one that could lead to censorship.

Finally, the CompuServe, *Phrack,* and Prodigy cases are only representative examples of some of the issues facing the computer-based communications field. As it continues to mature, there is a delicate balancing

act in the shaping of this emerging infrastructure. For example,

- How do you define free speech for these communication forms? What constitutes censorship?
- As was the case with CompuServe, who is liable for libel? Does anything go or should a company be allowed to control the dissemination of its information, much like traditional newspapers?
- Since they have different implications, should an electronic service be treated as a distributor or as a publisher in First Amendment issues?

These questions, among others, have been asked by writers, lawyers, and other individuals.[33] They are important also because their answers hold one of the keys to the future of our communications system.

The Democratization of Information II

The last statement leads us to a subject described in previous chapters, the democratization of information, and by extension, the potential birth of an electronic democracy. In essence, electronic communications conduits can be used to help ensure the free flow of information in a society and to provide individual citizens with a voice in their government. They can also be used to open up a dialogue between a government and its citizens, as stated in a 1990 report, *Critical Connections: Communication for the Future.* In fact, during a 1992 presidential primary, former governor Jerry Brown used two-way interactive services to communicate with voters during his campaign bid.[34]

At the same time, though, the emergence of this tool in the political arena has a number of implications. As was the case with the paperless society, will network access be limited to only those who can afford it? Or could the system be used to promote a specific political agenda rather than the free exchange of ideas?

To address these and other issues, there has been a call for establishing a cost-effective and national public network. It could be a video dial tone derivative, supported by cable, or possibly developed as a narrowband ISDN system. The latter proposal is the focus of our current discussion.

In brief, instead of puttering around on conventional telephone lines as we generally do now, this network, especially when combined with digital-compression techniques, would enable ordinary users to relay information ranging from computer data to video. But in the interest of making the network effective, in the sense of fulfilling its mandate of open access to millions of people, Mitchell Kapor, president of the *Electronic Frontier Foundation* (EFF), articulated several governing principles. According to Kapor, the proposed system, called the National Public Network (NPN), should

Ensure Competition in Local Exchange Services
    . . . Congress must act now to ensure competition in local exchange services. Competition will promote innovation in these services . . . and help guarantee equal access to all local exchange facilities.

Promote First Amendment Free Expression by Affirming the Principles of Common Carriage
    . . . [F]ull support for First Amendment values requires extension of the common carrier principle to these new media. Common carriage principles would require that public communications carriers offer their conduit services on a non-discriminatory basis, at a fair price, and interconnect with other communications carriers.

Preserve and Enhance Equitable Access to Communications Media
    The principle of equitable, universal access to basic services is an integral part of today's . . . telephone network. We must ensure that all Americans have access to the growing information services market now and in the future.[35]

There was also support for protecting privacy, using this system as the precursor for a future, high-speed network, and for making the system easy to use.

The latter point is especially critical. As discussed, one of the factors for the adoption of a given technology and service, depending upon the targeted user group, may be its ease of use. If a network is universally available but difficult to use, the person the system was designed to serve may not actually use it.

In sum, the NPN or similar system would take advantage of the existing infrastructure and the ongoing modernization of the telephone industry's physical plant to create a national information network. While it wouldn't equal the capacity of proposed high-speed networks, it could provide millions of people with easy access to a system that far exceeds current capabilities.

Applications and Implications. A system such as the NPN would be a communications conduit. It could support services ranging from exchanging E-mail to enabling people to retrieve information stored in a library across the country. It could also accommodate new information vendors, function as an electronic-publishing platform, and spark an economic revolution based upon new information services.

The network could also help create *virtual communities*, groups of people who may live in different regions of the country but share a common interest.[36] In this light, the NPN would extend the concept of a virtual community to more people, in contrast to the somewhat limited contemporary structure. Politically, it could foster the development of national electronic town halls or forums.

The network could also serve as an educational resource, and, by being widely available, would echo one of Thomas Jefferson's sentiments that "no republic could maintain itself in strength without the broad education of its people."[37] An NPN-type network could be a tool toward this end.

In a similar vein, it could complement the high-speed National Research and Education Network, described in Chapter 3. An important difference, though, for our discussion, is that the NPN can be viewed as another grassroots system designed to make an information network widely available to people, much like today's telephone.

As part of the overall communications sys-

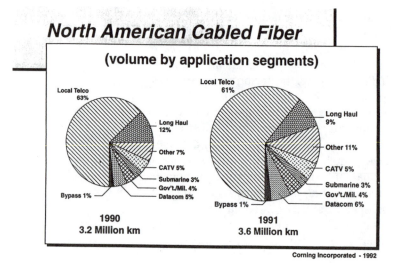

North American Cabled Fiber

(volume by application segments)

Local Telco 63%
Long Haul 12%
Other 7%
CATV 5%
Submarine 3%
Gov't./Mil. 4%
Datacom 5%
Bypass 1%

1990
3.2 Million km

Local Telco 61%
Long Haul 9%
Other 11%
CATV 5%
Submarine 3%
Gov't./Mil. 4%
Datacom 6%
Bypass 1%

1991
3.6 Million km

Corning Incorporated - 1992

Figure 10–7
Fiber-optic systems may provide the communications conduit for enhanced relays, ranging from video to data. This chart depicts the current user split. (Courtesy of Corning, Inc.)

tem, this type of network could also play a role in sustaining and encouraging the free flow of information in society. This is especially true when viewed in context with desktop publishing, desktop video, and other related technologies and applications.

Privacy must also be ensured, and as stated by Mitchell Kapor, First Amendment rights must be extended and guaranteed. Without such legal mandates, the network would be open to eavesdropping, censorship, and as with the CompuServe case, a potential chilling effect, which could lead to de facto censorship.

As a final thought, it's important to remember that technology is not or should not be the central issue in this scenario. Technological problems can be solved, and new developments may offer innovative solutions. For example, by using compression techniques, a telephone line could support relays ranging from voice to video to data. By extension, this ubiquitous link could provide millions of Americans with ready access to information resources and electronic communities scattered across the country and possibly even the world.

But unless the system is cost-effective and the interface is easy to use, its promise may not be fulfilled.[38] It may also have to become a part of everyday life, much like the telephone, for people to see the need for its existence.

REFERENCES/NOTES

1. Ronald J. Goldman, "KCET and Broadcast Teletext: Why We Came and What We've Done," *Communication Options*, April–May 1983, III, C1.
2. Joseph Roizen, "Teletext in the USA," *SMPTE Journal* 90 (July 1981): 603.
3. Actually, all the lines in a television frame do not contain picture information. Teletext magazines are inserted in a portion of the approximately 21 "empty" lines, the VBI, of a television signal.
4. Keycom Electronic Publishing and Southern Satellite Systems, *Technical Bulletin: Operation of Keyfax.*
5. The British had experimented with a system in which two related stories were, in a sense, linked. This system essentially eliminated the normal recycling time the system would go through to access a different page, in this case, the new story.
6. Even though it has been a few years, Electra was, for example, set up for less than $200,000. The low starting cost could prompt a local station to start its own teletext operation if a sufficient number of either television sets with internal teletext decoders or stand-alone decoders are manufactured and purchased. See Bebe F. McClain, "Taft Airs Affordable Teletext," *Broadcast Engineering* 25 (November 1983): 114.
7. Prestel, press release, August 29, 1983.
8. "The Electronic Mall Arrives; All Under One Roof (Yours)," *Viewtron Magazine and Guide* 1 (November 1983): 6.
9. "Knight-Ridder Pulls Plug on Viewtron," *Broadcasting*, March 24, 1986, 45.
10. C. D. O'Brien, H. G. Brown, J. C. Smirle, et al., *Telidon Videotex Presentation Level Protocol: Augmented Picture Description Instructions: CRC Technical Note 709-E*, (Ottawa, Canada: Department of Communications, 1982), 1.
11. Andrej Tenne-Sens, "Telidon Graphics and Library Applications," *Information Technology and Libraries* 1 (June 1982): 101.
12. Keith Y. Chang, "Videotex: A Pillar of the Information Society," paper, Department of Communications, Canada.
13. This is due, in part, to Prodigy's user interface and the service options it has generally supported.
14. Prodigy has also charged additional fees for specific optional services, including electronic mail beyond a set number of messages.
15. Shareware is software that you initially try

for free. If you like it, you then become a registered owner by sending the author a specified fee.

16. Dennis Fowler, "BBS of the Month," *Computer Shopper* 12 (August 1992): 705. Please note: If you are interested in linking up with a system, *Computer Shopper* has featured an extensive and updated BBS list.

17. Bitnet stands for "Because It's Time Network." See Tracy LaQuey, "Networks for Academics," *Academic Computing*, November 1989, 32, for more information.

18. Space Digest, Bitnet, v. 15, #127, August 20, 1992.

19. Real time implies, in this context, that the conversation is instantaneous. It is the written equivalent of the telephone. Also see Brendan P. Kehoe, *Zen and the Art of the Internet*, rev. 1.0, February 2, 1991. Besides serving as a guide to the Internet, it covers the etiquette of communicating on this type of system, such as the use of CAPITAL letters to convey a specific tone of voice.

20. Dennis Allen, "Editorial: Ethics of Electronic Information," *Byte* 17 (August 1992): 10.

21. Glenn Rifkin, "Do Employees Have a Right to Electronic Privacy?" *The New York Times*, December 8, 1991, section 3, 8.

22. ABC News, "FBI Pushing for Enhanced Wiretap Powers," transcript from "Nightline," show #2870 (May 22, 1992), 3. Please note: The collected information, such as a telephone conversation, could subsequently be used as evidence in court.

23. Sam Whitmore, "Drop a Dime and Stop Some Spooky Legislation," *PC Week* 9 (May 18, 1992): 100.

24. ABC News, "FBI Pushing for Enhanced Wiretap Powers," 2. The actual act was the Organized Crime Control and Safe Streets Act.

25. *Ibid.*, 4. This stance also appears contrary to the NTIA's assessment of the government's role in establishing communications standards, as described in Chapter 2.

26. J. H. Green, "Modem Wars," *Online Today* 2 (December 1983): 21.

27. Another proposal was in fact introduced in the early 1990s, where local line costs for service providers could substantially increase if they engaged in new service arrangements as may be provided by, for example, the ISDN. Basically, service providers could be locked into the current infrastructure unless they are willing to pay higher access fees, which in turn could be passed on to consumers. See CompuServe, *FCCPCS.TXT*, text file,

Library 2, Telecommunications Forum, or as reported, 6 FCC RCD 4524, for more information.

28. Barton Crockett, "DA Probes BBS Practices at Prodigy," *Network World* 8 (April 15, 1991): 4. Please note: Prodigy has also been accused of screening messages and for using an uneven editorial policy in determining what messages should or should not be posted on its bulletin boards.

29. Howard Rheingold, "The Thought Police on Patrol," *Publish* 6 (July 1991): 46. Please note: The document was concerned with the 911 emergency system.

30. Mitchell Kapor, "Why Defend Hackers?" *Effector* 1 (March 1991): 1.

31. *Cubby Inc. v. CompuServe* Inc., 19 Med. L. Rptr., 1528.

32. Lisa Picarille, "BBS Not Liable for Libel, Court Says," *InfoWorld* 13 (November 11, 1991): 130. Please note: The article also indicates that this particular case did not release the individual who ran the forum where the information was posted.

33. Please see Mike Godwin, *The First Amendment in Cyberspace* (Cambridge, Mass.: EFF, 1991), for more information.

34. Steve Higgins, "'Electronic Democracy' Wins Votes," *PC Week* 9 (May 18, 1992): 19. The use of new technologies in politics is not a new idea. See Robert G. Meadow, ed., *New Communication Technologies in Politics* (Washington, D.C.: The Washington Program of the Annenberg School of Communications, 1985), about such applications.

35. Testimony of Mitchell Kapor, president, the Electronic Frontier Foundation, before the House Subcommittee on Telecommunications and Finance, October 24, 1991. Excerpts.

36. U.S. Congress, Office of Technology Assessment, *Critical Connections: Communication for the Future* (Washington, D.C.: U.S. Government Printing Office, 1990), 190.

37. Noble E. Cunningham, Jr., *In Pursuit of Reason: The Life of Thomas Jefferson* (Baton Rouge, La.: Louisiana State University Press, 1987), 337.

38. The cost factor also includes consumer equipment.

ADDITIONAL READINGS

Burry, Yvonne H. "Calculating Your Risks." *Online Today* 5 (March 1986): 18–22. A sample of the applications and participants in the demographics field and how to obtain such on-line information.

Cate, Fred H., ed. *Visions of the First Amendment for a New Millennium*. Washington, D.C.: The Annenberg Washington Program, 1992. A compilation of articles that provides an analysis of First Amendment issues. One section covers the First Amendment and regulatory issues presented by new and converging technologies.

Connelly, Terry. "Teletext Enhances WKRC's Local News Image." *Television/Broadcast Communications*. October 1983, 52–58. An overview of the establishment of Taft Broadcasting's, now Great American Communications', *Electra* magazine.

Glossbrenner, Alfred. *Personal Computer Communications*. New York: St. Martin's Press, 1985. An excellent source book of PC-based and dedicated information systems. The topics include, among a long list of subjects, how to purchase and set up a modem, descriptions of the various systems, and hints on how to communicate with a system more effectively and time-efficiently.

KCET-TV. *Now! The Electronic Magazine: The KCET Teletext Project*. Los Angeles: KCET-TV. A comprehensive description of the development of an educational teletext project. An excellent guide for those who are interested in learning how such a project is established.

Kening, Dan. "A Connected Electorate." *CompuServe Magazine* 11 (September 1992): 35. The article provides an overview of some of the political applications of two-way interactive services. In one case, a CompuServe forum served as an information clearinghouse about the 1992 presidential candidates.

Martin, James. *Viewdata and the Information Society*. New York: Prentice Hall, 1982. An excellent source book about the technical elements of interactive systems as well as the applications and implications of their services.

McCabe, Kathryn. "Step Up to the Modem: Baseball Goes Online." *Online Access* 7 (Spring 1992): 14–15, 17. This article describes a baseball lover's dream come true: You can use an online service to manage a team you assemble while using the statistics from real players to keep track of your record.

Morgenstern, Barbara, and Michael Mirabito. "Educational Applications of the Keyfax Teletext Service." *Educational Technology* 8 (August 1984): 46–47. Examines some of the applications of a teletext magazine in an educational setting.

Ojala, Marydee. "Market Research: The Online Connection." *Online Access* 7 (Spring 1992): 18–21. The article covers how on-line services can be used to conduct market research.

Rayers, D. J. "The UK Teletext Standard for Telesoftware Transmissions." In *International Conference on Telesoftware*. London: The Institution of Electronic and Radio Engineers, 1984. Explores the technical issues of a telesoftware operation, using a teletext magazine, in this case, to distribute computer software.

U.S. Congress, Office of Technology Assessment. *Critical Connections: Communication for the Future*. Washington, D.C.: U.S. Government Printing Office, 1990.

Chapter 6 of the report explores "Communication and the Democratic Process." The topics range from the use of remote-sensing satellites to computerized information systems and BBSs.

GLOSSARY

Alphamosaic: The graphics-creation system originally adopted by the British for their teletext configuration. Alphamosaic graphics are very boxlike in appearance.

Alphageometric: The system that produces graphics that are more smoothly curved than their alphamosaic counterparts.

Computer forum: An electronic meeting place supported by a PC-based system. A forum supports a special interest group (SIG), such as individuals who own a specific brand of computer or who may belong to the same profession.

Decoder: The device that strips the teletext information from the VBI and displays the alphanumeric and graphics information on the television screen.

Dedicated information-retrieval system: A two-way interactive system that supports very specialized databases of information and, consequently, narrow and select subscriber groups.

Electronic bulletin board system (BBS): A miniature interactive system, usually comprising a PC, a hard drive, and special communications software. BBSs are established by, among others, government agencies, manufacturers, and private individuals.

Electronic mailbox: Most computer-based communication/information systems support an electronic mailbox, a portion of the computer system reserved for an individual subscriber's E-mail.

Gateway: A gateway, for our purposes, is an electronic gate that enables a subscriber to gain access to different and normally incompatible and inaccessible external computer systems.

Information provider (IP): An organization that provides information carried by an interactive service.

Interactive system: When applied to the communications and information systems described in this chapter, this term implies that an individual can literally interact with a given system, in this case, to retrieve or relay specific information.

Legal issues: There are a variety of legal issues raised by computer-based communications systems. Examples include privacy and content questions as well as the democratization of information.

National Public Network (NPN): A proposed computer-based communications system that most Americans would be able to use.

North American Broadcast Teletext Standard (NABTS): A teletext standard.

North American Presentation Level Protocol Syntax (NAPLPS): A videotex standard.

PC-based interactive system: A two-way interactive service that is tailored for PC users, including hobbyists and individuals who use PCs at work. The supported features range from E-mail to access to numerous databases.

Teletext magazine: An electronic magazine that appears on a television screen. Like a newspaper or a newsmagazine, the typical teletext magazine is composed of stories that range from international news to sports and weather information.

Vertical blanking interval (VBI): The component of the television scanning process that makes it possible to insert teletext information in a conventional television signal. Thus, when the television signal is distributed, the teletext information is likewise relayed.

Videotex system: A videotex system is a dedicated and interactive information-retrieval service. Videotex services have also been generally geared toward consumers, and graphics have been frequently used to represent and present information.

World System Teletext (WST): A teletext standard.

11 Teleconferencing

Chapter 10 outlined some of the developments in the computer-based communications field, including the launching of commercial interactive services and the proliferation of grass-root BBSs. The common thread running through all the systems is the exchange of information.

This chapter examines another type of information exchange, the *teleconference*. The word *teleconference* is an umbrella term that describes an electronic link or meeting between two or more locations. These meetings range from *audioconferences*, where people at different sites can hold two-way or interactive conversations, to *videoconferences*, where video information, pictures, can also be exchanged.

In addition to exploring teleconferencing, the chapter touches on two related topics, *computerconferencing* and *bypass* technologies and applications. Computerconferencing is an extension of the systems presented in Chapter 10 and is placed in this chapter, in part, for organizational purposes. In the context of our discussion, the focus is on the technology's electronic meeting characteristics.

Bypass systems are included in this chapter for much the same reason, organizational purposes. They can, for example, support a teleconference, even though this is not their primary function.

More important, bypass technologies and applications provide additional communication options. We now have the tools to communicate more efficiently and effectively, and we can break the physical constraints imposed by distance. Teleconferences accomplish much the same goal. In this respect, a bypass provides the means while a teleconference is one of the end results.

TELECONFERENCING: AN INTRODUCTION

As previously described, a form of teleconferencing is the videoconference. In a *two-way videoconference*, a popular configuration, the different parties can hear and see each other. In another variation, a *one-way videoconference*, the information exchange is a one-way audio and video delivery system. But there is an option for some interaction through a telephone or possibly a fax connection. In a typical example, a party receiving the information could ask a question by telephone.[1]

Videoconferences can also be either *motion* or *nonmotion* operations. The term motion implies that people can appear on the television screen in a lifelike manner. This capability, however, varies widely. It can range from full motion, where the transmission may look like conventional broadcast television, to limited motion, where movements may be jerky and there is a deterioration in the picture's quality.

A nonmotion or freeze-frame videoconference, the second category, consists of a series of still images that appear on the screen. Even though the visual element is not lifelike, an audio hookup could support a conversation.

Finally, a teleconference can be held over a range of communications channels, including satellite as well as FO and conventional telephone lines. A company may also own a private teleconferencing network, or as has been prevalent in the videoconferencing field, a public room can be rented for the occasional meeting.

The Codec. One of the devices that actually helped spur the growth of the telecon-

Figure 11-1
An example of a codec. (Courtesy of Compression Labs, Inc.; Rembrandt IIVP)

ferencing field, two-way videoconferencing in particular, is the *codec*. The term *codec* is a hybrid of the words *coding* and *decoding*. A codec digitizes and compresses a signal, that is, reduces its bandwidth requirement, so the information can be placed on lower-capacity and less-expensive communication channels. A codec at the other end of the relay reverses this process.

Compression has been discussed elsewhere in the book, in compression and FO lines, in desktop video applications, and for satellite relays. The teleconferencing field has similarly benefited from compression technology, and there is a wide range of codecs for this particular application.

In one case, a codec's final output could be a 1.544 megabits/second rate, in contrast with an approximately 90 megabits/second information stream that is produced when a broadcast-quality color signal is initially digitized. The transmission is also compatible with the T1 line; this is an important factor, since the signal must conform to existing transmission standards and systems.

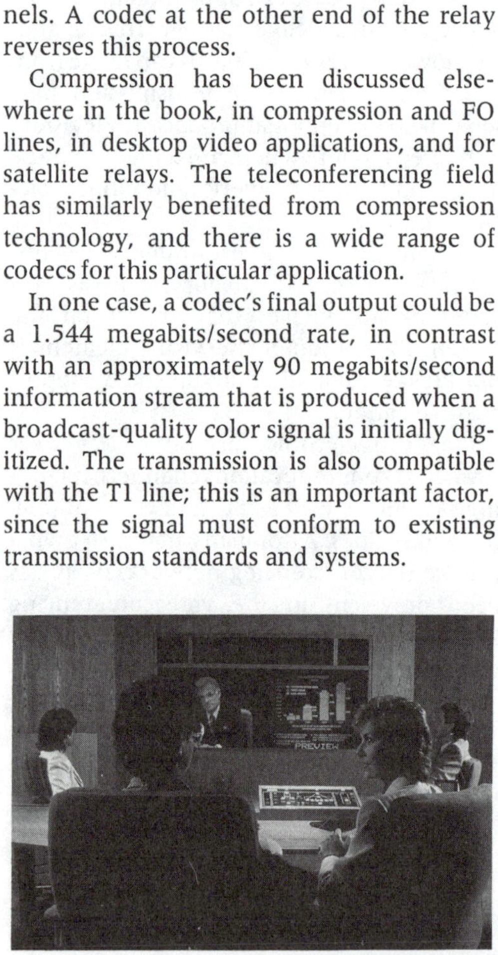

A codec may also support different standards with a variable transmission capability, including the relatively cost-effective 56 kilobits/second digital telephone standard. Other supported rates include those lower and higher than the 56 kilobits/second and 1.544 megabits/second transmissions, respectively.

As a result of the compression technique, though, a transmission's quality on a television screen may be inferior to one produced during a conventional television broadcast. This is especially true at the lower transmission rates.

In one example, small picture changes or motions can be accommodated, but rapidly gesturing hands and similar movements may not be smoothly reproduced. Thus, there is a tradeoff between reduced transmission speeds and costs versus the picture quality.

Teleconferencing: Specifics

This section examines, in a little more detail, different teleconferencing configurations. The first is a digital two-way operation.

Two-Way Videoconference. In a two-way videoconference, the various parties can see and hear each other, providing an interactive environment and exchange.

Two-way videoconferences have generally been used for *point-to-point* meetings, uniting two sites. But new advances have made it possible to take advantage of a multisite capability.

Regardless of its form, a two-way videoconference's primary advantage lies in its replication of a face-to-face meeting. The participants, also known as conferees, can react to each other's body language, valuable visual clues in interpersonal transactions.

A two-way videoconference can also accommodate a range of information, including graphics and possibly scanned documents. A system that fully integrates these capabilities, and a realistic visual representation of the videoconferencing participants, can improve a meeting's overall effectiveness.

To help facilitate this operation, the typical room can be designed with an electronic meeting in mind. For example, the table where the conferees sit can be shaped to

Figure 11-2
A teleconferencing setup. Note participants in this room (foreground) and the shot of a conferee at another location (background screen). (Courtesy of US Sprint)

maximize their view of the monitors in the room. The goal is to promote good eye contact between the parties at the different sites.[2]

Strategically placed videocameras can also produce various shots of the conferees, while an additional camera can shoot graphics or other supporting materials. Proper lighting and acoustics similarly play key roles in this operation.

Before the development of such dedicated videoconferencing rooms, some facilities were hampered by inadequate equipment and design considerations. A large office may have been converted into a videoconference room by simply adding the necessary communications equipment. Thus, lighting and other production considerations, the human element that would have made people more comfortable in this environment, may have been overlooked.

Another past and current problem has to do with the notion of appearing on camera. Some people get nervous; others may think that a videoconference is an unnatural way to communicate when compared to an in-person, face-to-face meeting.

One-Way Videoconference. The one-way videoconference, which has also been called business television, is typically relayed via satellite from one location to a number of geographically scattered sites (*point-to-multipoint*). A one-way videoconference has generally been analog in nature, but recently available digital-compression techniques, when combined with satellite delivery, can make for an effective relay. In this situation, though, the picture quality is of paramount concern, in contrast to a two-way videoconference, where a lower image quality but a higher compression ratio may be more acceptable.[3]

In a typical application, the audio and video information can be a one-way stream from a corporate headquarters to its branch offices. These sites may communicate with headquarters, in turn, with a telephone or other audio-based hookup. This connection can be used for question-and-answer periods, the discussion of relevant points, and for the clarification of specific details.

In reflection of its point-to-multipoint ca-pabilities, the one-way videoconference has led to the development of dedicated video-conferencing networks. One may support educational applications, while another may function as a corporation's private network.[4]

Besides these networks, one-way video-conferencing has been supported by organizations that lease satellite time and the requisite facilities. One-time or occasional video-conferences, also called special events or ad hoc videoconferences, have also been accommodated.

Finally, the power of this type of electronic meeting has been vividly demonstrated in medical videoconferences. A doctor can, for example, perform a cornea transplant, and this specialized surgical procedure can be relayed by satellite to other locations. Thus, a videoconference would make it possible for a number of doctors to view and learn a valuable medical technique through this real-time meeting.

Nonmotion Videoconference. In a non-motion videoconference, still images are delivered over voice-grade or faster communications channels. The actual relay time can vary, depending on the communications line and the image's resolution, and either dedicated or PC-based configurations can be used.

A nonmotion videoconference has a number of advantages as a visual communications tool. A system can be relatively inexpensive to set up, and the transmission cost could be reduced to the level of a local or long distance telephone call. A motion videoconference, in contrast, typically has higher equipment costs and is relayed on wider and more expensive channels.

With regard to applications, nonmotion videoconferences are suitable for those situations where visual information must be exchanged, but not necessarily moving pictures. These have included X rays and illustrations for the medical and educational fields.

Audioconference. An audioconference can be thought of as an extended telephone conversation, but instead of talking with only one person, you may be talking with several or more people. In such an operation, multiple sites can be connected through a teleconferencing bridge.

An audioconference is a satisfactory communications tool in many situations, and it is relatively inexpensive to implement. Like videoconferencing configurations, audioconferencing is also supported by its own array of equipment, including special microphones that help facilitate and enhance the operation. In one example, a microphone may have a 360-degree pickup pattern so it can serve several or more people sitting at a table.

An audioconferencing system can also be used in various applications. It can create an inexpensive communications link between a journalism class and a sports reporter, who may live and work 200 miles away from the college. It can also support business meetings and individuals engaged in a national political campaign.

The primary criterion for adopting audioconferencing or one of the visual systems is an organization's specific needs. In many cases an audioconferencing link may suffice.

Other Considerations

A teleconference can take many shapes and support numerous applications. Several complementary developments also helped boost the industry's overall growth. In fact, it experienced a surge in the 1990s when sales of products and services for the videoconferencing sector alone rose from $127 million to $510 million in 1987 and 1991, respectively. This figure is expected to reach $1.5 billion during the mid-1990s.[5]

Some of the reasons for this growth include the following.

Standards. A standard, H.261, also known as Px64, has brought a level of compatibility between different codec manufacturers.[6] It is one of a family of standards that falls under H.320, and while it's not the first attempt at standardization, it does allow different codecs to work with each other.

Transmissions. The industry has benefited from lower transmission and equipment costs. It's less expensive to hold a videoconference, and the quality has improved. In this regard, newer codecs can support an enhanced relay at lower transmission speeds.[7]

It's also easier to hold an electronic meeting. Digital dial-up lines, which can make a relay essentially as common and transparent to the user as a telephone call, are becoming widely available. This situation will only improve over time as digital systems continue to proliferate. These developments may even include, as previously outlined, using the current telephone infrastructure for the delivery of sophisticated information.

Flexibility. Teleconferences are becoming more flexible in the range of information they can accommodate. One example is the multimedia videoconference, where computer capabilities have been married with those of a conventional videoconference. In a typical operation, graphics and computer data can be exchanged, documents can be

Figure 11–3
A teleconferencing system that can support a variety of options including videoconferencing, audioconferencing, document exchange, and computerconferencing. As indicated in the text, the capability to exchange a range of information enhances the communications process. (Courtesy of VideoTelecom Corp.; Benchmark 225)

electronically annotated, hard copies can be printed, and customized and simplified user interfaces can be set up.[8] The basic idea is to provide users with all the tools they might use during a conventional face-to-face meeting.

Besides this development, teleconferencing in general is becoming more flexible as the industry matures. Organizations can rent public rooms to experiment with videoconferencing or for occasional use, and turnkey systems, which may include both hardware and the necessary communications links, are supported. Satellite and FO relays are now widely available, and networks such as the Sprint Meeting Channel have accommodated domestic and international relays.[9]

A company can also use a portable and less expensive rollabout to hold a videoconference. This unit can be moved from room to room and eliminates the need for a dedicated videoconferencing facility. However, its aesthetic elements could be enhanced if the unit is used in a board or conference room where lighting and acoustics have been taken into consideration.

Personal Videoconferencing. Videoconferencing has also come to the desktop. In the early 1990s, a number of desktop configurations were introduced on the market. These devices can trace their origins to AT&T, which experimented with the Picturephone transmission standard and the Picturephone, its hardware component.

The Picturephone was a compact desktop unit that incorporated a small television monitor, camera, and the necessary audio components to establish a link with another Picturephone user. Even though the service was, in a sense, ahead of its time, it did not catch on in either the business or private sectors. The transmission and hardware costs, a Picturephone user's possible reluctance to appear on a television screen, and the inadequate reproduction of printed documents hampered sales.

Since 1964, when AT&T publicly demonstrated a video telephone at the World's Fair, newer products have been introduced.[10] These include consumer-oriented nonmotion units, which relayed still black-and-white images over standard telephone lines, and newer motion systems.

The latter have ranged from PC-based configurations to stand-alone desktop models; both analog and digital transmission schemes have been supported. The targeted user group includes consumers and businesses, and the costs and capabilities of the different outfits have varied.

While personal videoconferencing may be inevitable, the time frame for its broad acceptance is unknown. Standards are not finalized for this particular application area, and the quality, as of this writing, may still be unacceptable to some users.[11]

Another, and possibly more important, factor is the human element. New transmission lines, as outlined in previous chapters, may support enhanced relays, and equipment costs may continue to fall. But systems geared for the consumer may run into the same problem faced by the original Picturephone: many people may not want to appear on camera. Think of your own use of the telephone. When you talk with someone, are you always attentive? Or depending upon the conversation, are you simultaneously working on a computer, eating, watching TV, or as indicated in research that evaluated the Picturephone, reading, doodling, or even yawning?[12] While you could use a system in a voice-only mode, would the person you're talking to be insulted?

Consequently, this type of telephone system calls for a different mindset on the part of the user. It may be fully accepted only when a generation of children actually grow up with this technology and it is a part of their everyday lives. The introduction and widespread use of new processors, which may help turn the ubiquitous PC into a ubiquitous videoconferencing console, may help in this regard.

Advantages of Teleconferencing

Now that the teleconferencing field has been covered, it may be appropriate to examine some of the advantages offered by teleconferencing, regardless of its form.

One advantage is the capability to enhance

an organization's level of productivity. A teleconference can link an organization's branch offices and can facilitate the flow of information between these sites, either as needed or on a regular basis. With its own private setup, a company could also meet on short notice to respond to crisis situations.[13]

Another advantage has to do with resources. In the educational field, for instance, teleconferences have been arranged where guest lecturers, who may live too far away for a campus visit, are able to "electronically" meet with a class. In this application, a teleconference makes it possible for an organization to draw upon resources, an expert in a specific field, who may not be available under normal circumstances. Businesses can be similarly served.

Employee morale can likewise be enhanced if management uses a teleconference to keep employees informed about important and relevant news items. Time and money can also be saved by reducing travel between sites.

In the latter case, an individual's job may call for extensive traveling, which normally entails a drive to the airport, waiting and flight time, another drive to the meeting point, and the return flight. This itinerary may then be repeated for other sites. In today's world, this wasted time and energy are the two resources that can never be recouped. The financial cost can also be substantial.

Teleconferences were designed, in part, to alleviate these problems. If an electronic meeting could be arranged between offices on the east and the west coasts, the same information could be exchanged through a teleconference, rather than a series of trips.

But teleconferencing will not eliminate all travel. Interpersonal communication is important, and face-to-face meetings are still called for on different occasions. For example, if people are learning how to use a new piece of complicated equipment, which requires an extensive manual setup, an on-site seminar may be appropriate. In this setting, the instructor can actually sit down with these individuals and, if necessary, make adjustments. But once this initial phase is completed, a teleconference, or even a telephone call, may be suitable for follow-up work. In another situation, some people still prefer, as previously indicated, in-person meetings.

Nevertheless, teleconferencing is apparently emerging as another tool that organizations, and possibly individuals at some future date, can use to improve their level or quality of communication. All the advantages brought about by teleconferencing can also be extended to the international market. Since many companies are now global in nature, a teleconferencing operation can help unite these geographically distant sites. This capability cannot be overemphasized, as international considerations continue to play an integral role in domestic operations.

Conclusion

The teleconference is an important element of our overall communications system and has emerged as a valuable tool: it can save time and help you to use your time more efficiently.

While many businesses are now taking advantage of this technology and its applications, an electronic meeting may still be viewed, in some quarters of society, as a special and out of the ordinary communications event. But the development of PC-based and inexpensive videoconferencing outfits, which may become standard fixtures for both the office and home, may alter this perception. Sophisticated and cost-effective communications channels, enhanced display systems as promised by HDTV technology, more powerful processors, and other factors should only accelerate this trend.

Figure 11–4
One element that may boost teleconferencing is the introduction of more transparent and easier-to-use control interfaces, such as the one depicted in the photo. (Courtesy of VideoTelecom Corp.)

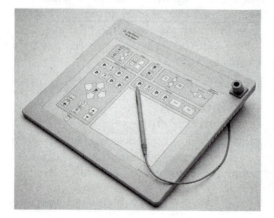

Similarly, children who have grown up with the technologies that are the driving force behind this field may accept the teleconference as a natural part of their lives. They may already be computer literate, and thanks to the proliferation of home videotape systems and a new generation of electronic toys, appearing on camera will just be old hat for these individuals.

COMPUTER-CONFERENCING

The term *computerconferencing* can describe a computer-based meeting. For our discussion, a conference can range from an exchange of pictures between two people to dedicated networks that may link a number of domestic and international users.

In one application, a PC-based configuration could unite multiple sites, as would be the case with a national real-estate company. Pictures of different houses could be relayed through this communications system, and an office in Denver could send a series of pictures to the New York branch for review by a client who is moving out west. A telephone conversation could supplement this exchange.

Some of the computer-based systems described in Chapter 10 can also support conferencing. An advantage of this type of operation is its ready availability. Groups of people scattered in different geographical locations can easily gain access to the system. The company also provides the organizational and communications infrastructures that may be required to hold the meeting.

In a different setting, a business or educational institution may opt to create its own internal network. Meetings could be rapidly convened and a variety of information could be exchanged. The conference could be viewed, in this light, as an interactive extension of E-mail.

Special hardware and software systems have also been developed to support computerconferencing activities in both real time and non–real time. Real time, in this context, implies that messages can be sent and received as you view the screen and interact

with the system and the other people tied in the network. The non–real-time elements, in contrast, may encompass a series of longer messages, a central database of information, and a record of current and past comments that all the conferees can see.

Telecommuting and the Electronic University. On a smaller scale, computerconferencing can be held between a single individual and an organization, as in the case with a *telecommuter*. A telecommuter works at home and maintains contact with the office by computer. This individual, who may make only a set number of personal visits to the office, can be an employee, a consultant, or a chief executive officer.

Telecommuting is becoming a popular work option. It can, for example, save travel time, especially if you work in a high-density urban area where commuting on an overextended highway and mass transit system may take an hour or more. Telecommuting also gives you more options, since you can choose to live farther away from your place of employment, or work at a job that might normally be geographically inaccessible.

Besides these advantages, telecommuters may be more productive and experience a higher level of job satisfaction than their office-bound co-workers. There are also corresponding social benefits, such as fuel savings, air-pollution reduction, and improvements in child care, since a parent can now work at home.

Several factors have combined, at this time, to make telecommuting a viable work alternative. They include the widespread adoption of the PC and, as described throughout the book, the vital role that information, typically generated and manipulated by computers, plays in most organizations.

Yet not everyone is happy with this state of affairs. There is the potential for abuse, where an individual may be treated as a contract worker instead of a salaried employee with all the attendant benefits. There is also the potential for lower wages, since a worker may be cut off from the "informal communication channels and . . . less well integrated into an organization's structure and culture."[14] An individual may also be uncomfortable with this work environment and prefer the interpersonal contact a conventional job may afford.

On a broader scale, telecommuting could reduce human communication to a machine-dominated format. This could, in turn, have an adverse effect upon the socialization process, since the interpersonal relationships we develop at work may play an important role in our lives.

Isaac Asimov, the late science-fiction writer and science authority, paints an interesting portrait of a society in which communications technologies have contributed to the development of a rigid social structure in which all forms of personal human contact are avoided. Even casual meetings are conducted through lifelike three-dimensional visual relays.[15]

Could our own society be affected, albeit to a lesser degree, by telecommuting? What if this factor is weighed with other technologies that will enable us to shop and bank at home, and to create sophisticated home entertainment centers?

Finally, while it is clear that telecommuting is not perfect, it does support a new and potentially time-efficient work environment. It will also be enhanced by high-speed digital lines as they become integrated in the overall communications structure. A high-speed line would enable the telecommuter to efficiently exchange a high volume of information, and in fact, this line could support videoconferences held with employees at other sites.

The concept of telecommuting has also been extended to the field of education, where individuals can register for computer-based and -delivered college courses. In a typical situation, a student registers for a course and is assigned a teacher, who in this case functions as an "electronic" instructor. The school also operates like any other educational institution. Textbooks are selected, lectures for the course can be downloaded and reviewed by the student, assignments are graded, and depending upon the system, the teacher may be available for consultation during specified electronic office hours or through E-mail.

This operation enables an individual who may live too far away from a college, or whose schedule may not coincide with a local institution's, to take a college course. The system could also accommodate an individualized study plan, is cost-effective, and could help a student enhance his or her critical-thinking skills through a varied course selection. If desired, it may even be possible to pursue a degree.

An educational system based upon this model may also be a portent of more sophisticated educational networks. As new forms of integrated communications lines are established, it may be possible to extend an electronic course so it encompasses sophisticated graphics that could complement a lecture. It may also be feasible to hold, on occasion, an interactive conference for the various scattered class members.

BYPASS SYSTEMS

A teleconference, be it one or two way in nature, can be conducted as a bypass operation, where an organization bypasses, or goes around, the traditional public telephone network.[16] Other, more common applications, include voice and data transmissions.

Thanks in part to the divestiture of AT&T in the early 1980s, a new dawn greeted the telecommunications world. Different segments of the industry that had been dominated by AT&T became open markets for large and small companies, which competed for their share of the sales and leasing opportunities.

While this development was taking place, individual companies became more responsible for their own communications systems. For example, various intra- and intercity links were established that could be more responsive to an organization's specific and changing communication needs. Thus, a company could more readily react to new communication demands since it used private or leased links. Just as important, this line could potentially provide the organization with a faster and higher-capacity communications channel than those supported by the established public systems.

Finally, for our discussion, there is another type of bypass, a *personal bypass*. These systems, which are still maturing, free us from physical constraints and, potentially, could allow us to communicate from any location in the world. Some of the more traditional and personal bypass systems include the following.

Microwave

Even though the microwave system has traditionally served as a long-distance communications link, the technology has been widely adopted for intracity and short-distance intercity operations. It is cost-effective and can accommodate a range of information with a wide channel capacity.

The technology may also be easier to implement than an FO- or copper-based system. You don't have the potential problem of obtaining the right-of-way clearance to lay the cable.

On a negative note, a microwave transmission may be affected by heavy rain. An FCC license must also be obtained to initiate an operation, and a microwave system is a line-of-sight medium. The transmitter and receiver must be in each other's line of view. This may necessitate, in urban areas, the placement of the components on the roofs of high buildings to avoid potential obstacles, such as other buildings, that may block the transmission.

Over-the-Air Laser

An infrared laser can be used to relay voice, video, and computer information through the air from one site to another, much like a microwave operation. The laser system is cost-effective, has a very wide channel capacity, and is a very secure medium. It is hard to tap or steal information due to the nature of the transmission.

On the negative side, a laser relay is also a line-of-sight operation and may be affected, to a varying degree, by smog and other atmospheric conditions. But unlike its microwave counterpart, an FCC license is not required. An operation can even be initiated in highly developed urban areas where it may be impossible to set up a microwave configu-

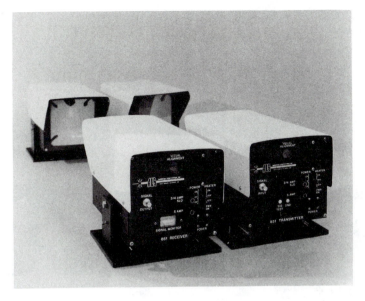

Figure 11–5
An over-the-air (atmospheric) optical communications system. This type of system can accommodate a range of information. (Courtesy of American Laser Systems, Inc.; Model 851)

ration if all the available frequencies have already been assigned.

Fiber-Optic and Copper Cable

FO and standard copper lines are used in bypass relays. Copper cable is the backbone of many communications systems, while an FO line has a greater channel capacity. Both systems, however, demand right-of-way clearances, which may be quite expensive to obtain, if at all, before the cables can be laid.

Cellular Telephone

A cellular telephone is one of the personal bypass systems. In this configuration, a specified geographical region is divided into small physical areas called *cells*, and each cell is equipped with a low-power transmission system to form the communications link. Since the cellular telephone is designed to support mobile activities, it allows an individual to use the system in a car, on the street, and in other locations.

In one example, as an individual approaches a cell's boundary while driving, the signal between the telephone and the transmitter becomes weaker. At this point, the telephone connection is basically picked up by the new cell the car is approaching. The phone is then switched over to and operates on a different frequency, to avoid potential

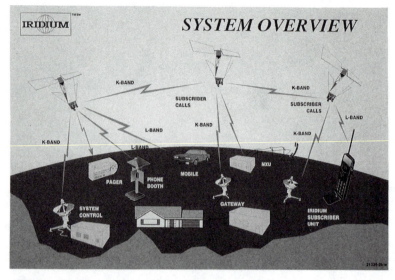

Figure 11-6

This proposed space- and Earth-based operation could support a sophisticated personal communications network. The various elements are highlighted in this diagram, including the satellites and personal communications device (subscriber unit is at the bottom right-hand corner). (Courtesy of Motorola, Inc.; Iridium)

interference with adjacent cells. This procedure is automatically completed by a sophisticated control network that helps maintain the communications link.

Cellular technology has also been integrated with portable PCs for remote data relays, and in a related development, digital operations may fuel the industry's growth. A digital link should support a cleaner signal and will expand the current network's capacity.[17]

The same digital technology may give an added bonus, privacy. A shortcoming of cellular technology, as of this writing, is the ability to intercept conversations and data relays. For example, by using a fairly inexpensive scanner, a supposedly private conversation may not be private anymore. This type of monitoring can invade an individual's privacy and could even lead to industrial espionage if information is illegally intercepted and used.[18]

But digital technology could provide some measure of security. A digital relay could make a conversation unintelligible to the average scanner owner. Encryption schemes could also be readily adopted for voice and data; but as described in Chapter 10, this level of security may be compromised through government regulation.

In sum, cellular technology has made it possible to relay information and to hold a conversation from the field. Cellular users have also benefited from technological ad-

vancements. Call-forwarding and other optional services are available, and you can select a telephone to match your specific needs. In this case, you may opt for a mobile phone installed in your car, if signal quality and range are your primary criteria, or you can use a portable phone that can literally go where you go.

Satellite Communications

Satellites are especially well suited for long-distance communication and point-to-multipoint operations. The very small aperture terminal (VSAT) has also emerged as a powerful element of the overall bypass market, and a new type of satellite system may lead to the creation of a global personal bypass.

The latter operation could be considered a space-based extension of cellular or mobile communications technology. It could also complement and supplement the services provided by geostationary communications satellites.

To actually create the system, a series of small satellites would be launched and placed in low Earth orbits (LEOs). This constellation of satellites could provide global coverage while their low altitudes would support portable transmitters and receivers on Earth.

In one potential application, short bursts of data, which may include messages from remote locations, could be relayed over small, inexpensive terminals.[19] In a proposal from Motorola, Inc., called *Iridium,* a multi-satellite network would function as an interconnected Earth- and space-based venture.[20] It would offer a range of services and would allow a user to talk with a friend half a world away through portable, mobile telephones. The same system could also serve as a relay platform for nations without a developed communications infrastructure.

All in all, it's an interesting concept, especially in light of the history of our satellite system. As you may recall, LEO satellites were abandoned in favor of their geostationary counterparts for communications applications. Yet the new generation of LEO satellites, when properly configured and coordinated, could deliver certain satellite services right into our hands. It's almost like Dick Tracy and the wrist radio come to life.

The Spectrum and the Personal Bypass

The development of an Iridium-type system would necessitate a spectrum allocation, and during a 1992 World Administrative Radio Conference, the allocations were actually granted.[21]

As indicated in Chapter 2, though, spectrum space is limited and valuable real estate, and different companies are competing for allocations. These include an emerging category of services that have been pioneered in Europe and have been variably called *personal communications networks* (PCNs) and *personal communications services* (PCSs). As defined by the FCC, a PCS is "a family of mobile or portable radio communications services" designed to meet the "communications requirements of people on the move."[22]

In an industry where names may change as frequently as the weather, the basic premise behind this class of operation, regardless of the name, remains the same: to deliver mobile, wireless communications services. As of this writing, it's also a wide-open market.

Computer-to-computer as well as voice relays could be supported. Both local-area and broader geographical communications links could be established, and ties could be created between the developing and existing communications networks. This emerging field could almost be viewed as a test bed for new personal bypass technologies and their applications.[23]

Like the satellite-based personal communications systems, PCNs and PCSs could also complement and supplement our current communications infrastructure. They may free you, for instance, from having to plug your computer or telephone into a wall outlet for certain communications applications. Similarly, a portable PC, or possibly a personal digital assistant as described in Chapter 3, will automatically receive E-mail and will allow you to send data back to your home or office. You may also be able to tap into a worldwide LEO network, dial a specific number, and reach another person via a portable phone, wherever he or she may be situated. It will almost be like super call-forwarding, free of the constraints imposed by wires, distance, and location.

But despite this field's promise, there may be losers in this scenario. The available spectrum space may be inadequate to support all requests, and preexisting services may have to be shifted due to the new assignments.

CONCLUSION

All the technologies and applications described in this chapter have contributed to the growth of the communications industry. When examined as a system, they compose a diverse collection of tools.

A videoconference, for instance, makes it possible to hold a face-to-face electronic meeting on short notice, while computer data can be used as an element of a multimedia videoconference. In another type of electronic meeting, computerconferencing can stand as a teleconferencing tool in its own right.

But perhaps the most flexible sets of tools are those offered by the standard and personal bypass technologies and applications. In one case, a constellation of LEO satellites could create a new mobile and global com-

Figure 11–7
This photo reveals the potential compact size of a personal communicator device that could tap into an LEO satellite communications network. (Courtesy of Orbital Sciences Corp.)

munications network. PCNs and PCSs, for their part, promise to deliver a host of new services.

These operations are also interdependent. One series of technological developments affects another. Two-way videoconferencing serves as an example.

A new generation of codecs, when combined with the widespread availability of suitable transmission lines at a reasonable cost, has contributed to this form of videoconferencing's popularity. Basically, technological developments have made it possible to produce a higher-quality relay, which in turn can be delivered on less-expensive and more widely accessible communications channels.

The potential launching of an LEO satellite network serves as another example. The technology now exists to build sophisticated small satellites that are matched by portable telephones and other communications devices. This entire system, in turn, is influenced by developments in the satellite-launch industry. Without cost-effective ELVs, the deployment of a constellation of satellites could be prohibitively expensive.

There are also a number of developing technologies and applications that could similarly affect this field, and vice versa. These include the possible delivery of motion video on telephone lines and the creation of an NPN-type network, as described in the previous chapter. Both operations could potentially support teleconferencing and would provide the necessary infrastructure to reach most homes.

Yet there are still some unanswered questions. Even if hardware prices drop and transmission costs are reasonable, would people, at this time, be willing to appear on camera? Another question touches on a different element of the privacy issue. Some individuals believe the new technologies may be too much of a good thing.

If a call could be routed to you wherever you are, and if your home becomes a workplace, what impact will this have on your individual privacy? Will the "electronic clutter" become so pervasive that these tools become more of a distraction than an aid?[24] In essence, is the ability to gain access to information at every waking moment, whether you're in a car or in an office, necessary? Or do we also require quiet, reflective periods, free from distractions? More pointedly, how do you strike a balance between the two?

Finally, this concept of delivering information to a home or business can be extended even further through *virtual reality*. As described in the book's Afterword, a virtual reality setup would bring Isaac Asimov's idea of three-dimensional meetings closer to an actual reality.

REFERENCES/NOTES

1. Bill Dunne, "Screening the Confusion Out of the Different Types of Videoconferencing," *Communications News,* November 1985, 34.
2. David B. Mensit and Bernard A. Wright, "Picturephone Meeting Service: The System," in *Teleconferencing and Electronic Communications: Applications, Technologies, and Human Factors* (Madison, Wis.: Regents of the University of Wisconsin, 1982), 174.
3. Clarke Bishop, Ken Leddick, and Jim Black, "Compressed Digital Video for Business TV Applications," *Communications News,* December 1991, 40.
4. The International Teleconferencing Association, "ITCA Teleconferencing Definitions," *Guide to Membership Services,* brochure.
5. The International Teleconferencing Association, "Business Television Private Network Market Reaches $606 Million," press release, February 1992, from a survey by TRI and IFC Resources, Ltd. of 100 companies responding to the impact of videoconferencing on their operations.
6. "Video Communications Comes of Age with H.261 Standard," *Communications News,* February 1991, 10.
7. Robert Lindstrom, "Shrinking of the Globe," *Presentation Products* 6 (June 1992): 22.
8. Conversation with VideoTelecom Corporation, Summer 1992, developer of such a teleconferencing system.
9. Sprint Communications, "Sprint Meeting Channel-Rates and Services," brochure.
10. Patrick Portway, "The Promise of Video Telephony Made in '64 Finally Is Being Fulfilled at Reasonable Prices," *Communications News,* February 1988, 35.
11. Elliot M. Gold, "Trends in Desktop Video and

Videophones," *Networking Management*, May 1992, 46.

12. Howard Falk, "Picturephone and Beyond," *IEEE Spectrum*, November 1973, 48.

13. The International Teleconferencing Association, "Business Television Private Network Market."

14. Robert E. Kraut, "Telecommuting: The Trade-Offs of Home Work," *Journal of Communication*, Summer 1989, 43.

15. Isaac Asimov, *The Naked Sun* (New York: Ballantine Books, 1957), 46.

16. Dwight B. Davis, "Making Sense of the Telecommunications Circus," *High Technology*, September 1985, 20. Please note: Companies have been using a bypass system within the physical confines of the organization for a number of years. The Private Branch Exchange, or PBX, serves as a privately owned switchboard of sorts primarily for a company's telephone operation. Thus, it bypasses the local telephone system for this type of internal hookup, and it is likewise the link to the outside world, the public telephone system. PBX systems, as well as other communications operations, were given a boost by the 1968 Carterphone decision. In essence, after this time, equipment manufactured by private companies could be connected with the public telephone system.

17. Anthony Ramirez, "Next for the Cellular Phone," *The New York Times*, March 15, 1992, Section 3, 7.

18. Robert Corn-Revere, "Cellular Phones: Only the Illusion of Privacy," *Network World* 6 (August 28, 1989): 36.

19. Orbital Sciences Corporation, "Orbcomm Signs Marketing Agreement in Five More Countries," news release.

20. Jim Foley, "Iridium: Key to Worldwide Cellular Communications," *Telecommunications* 25 (October 1991): 23. Please note: The ground-based sector would handle, in part, billing and the necessary authorization to use the system. This would be provided through a number of gateways.

21. Jack Messmer, "Int'l Decisions Reached on Allocating Radio Spectrum," *Network World* 9 (March 9, 1992): 2.

22. From an August 1992 *FCC Notice of Proposed Rule Making and Tentative Decision;* GEN Docket #90–314; downloaded from CompuServe, prepared by Scott Loftesness, August 25, 1992.

23. Conversation with FCC engineering department, September 16, 1992.

24. Patrick M. Reilly, "The Dark Side," *The Wall Street Journal*, November 16, 1992, technology supplement, "Going Portable," R12.

ADDITIONAL READINGS

Bernard, Josef. "Inside Cellular Telephone." *Radio-Electronics* 11 (September 1987): 53–55, 93. An overview of the technology and applications of this growing field.

Glass, Brett. "Bulletin Board Systems: A Primer for Business." *InfoWorld* 12 (January 8, 1990): supplement, S1–S5, S11. A primer about BBSs.

Gold, Elliot M. "PC Graphics Add a New Dimension." *Networking Management*. April 1992, 38–41. And, Halhed, Basil R., and Lynn D. Scott. "Making Multimedia Conferencing Work for You." *Video Systems* 17 (December 1991): supplement, 2–4. An examination of the use of graphics in videoconferencing and multimedia conferencing.

Gold, Elliot M. "Trends in Desktop Video and Videophones." *Networking Management*. May 1992, 44–48. As stated by the name, a look at desktop videophones.

Greelis, Michael, and William Veatch. "Audio Teleconferencing: A Simple, Frugal Alternative." *Biomedical Communications* 11 (August 1983): 24–27. An overview of the audioconferencing field. The article can also help individuals, in terms of equipment options, who are interested in conducting an audioconference.

Grotta, Daniel, and Sally Grotta. "Work Anywhere." *Home-Office Computing* 10 (March 1992): 39. An overview of portable PCs and cellular phones for working in the field.

Halhed, Basil R., and Lynn D. Scott. "Videoconferencing Terminology." *Video Systems* 18 (May 1992): supplement, 8–10. A list and accompanying definitions of the more important terms in the videoconferencing field.

Kames, A. J., and W. G. Heffron. "Human Factors Design of PICTUREPHONE Meeting Service." A paper presented at Globecom '82, sponsored by the IEEE, in Miami, Fla., November 29 to December 2, 1982. A review of an AT&T videoconference service. More pointedly, the paper discusses the various tests that were conducted to judge the system's effectiveness as a communications tool. Participants, for example, rated the quality of the system's audio and video signals.

Llana, Andres, Jr. "The Prospect for Personal Communications." *Telecommunications* 26 (September 1991): 37–40. An examination of the development of PCN/PCS operations and possible U.S. configurations.

Networking Management. Mid-November 1992. "Videoconferencing and Teleconferencing Directory."

This special edition is an invaluable guide to different teleconferencing forms and vendors.

Olgren, Christine. "Trends in the Use of Teleconferencing Illustrate the Wide Variety of Applications That It Offers." *Communication News,* February 1988, 22–26.

A very comprehensive and informative article compiled from a 10-year trend analysis of teleconferencing. It highlights applications, the users, and the growth of the different teleconferencing systems.

Rapaport, Matthew. "Computer Conferencing in the Corporate Environment: Will It Succeed?" *Telecommunications* 25 (May 1991): 23–27. An interesting look at why text-based internal computerconferencing has generally not been successful in the current corporate environment.

The Wall Street Journal. November 16, 1992. Technology Supplement: "Going Portable." This supplement covers a range of portable-computing and communications topics. Examples include Laurence Hooper, "No Compromises," R8; and "Getting Personal," 27, 29, an interview with John Sculley.

GLOSSARY

Audioconference: A form of a teleconference. In an audioconference, individuals at two or more sites can speak to and hear each other.

Bypass system: A private or leased communications system that can bypass the standard commercial and public communications facilities.

Cellular telephone: A communications tool and system based upon frequency reuse and a sophisticated monitoring design. An individual equipped with a cellular telephone could make a telephone call while driving, while on a boat, or even on the street. A system may also accommodate computer relays.

Codec: An hybrid of the words *coding* and *decoding.* A codec is a component that converts an analog signal into a digital format. In a teleconferencing environment, it also compresses the signal in the sense that the information can be relayed on lower-capacity and less-expensive communications channels. A codec at the end of the relay converts the signal back into an analog form.

Compression: Compression techniques are used in teleconferencing applications so signals can be relayed on lower-capacity and less-expensive channels.

Computerconference: A meeting conducted through the use of computers that can support the exchange of information ranging from text to graphics.

Electronic university: A school that offers courses via computer.

Motion videoconference: A motion videoconference can duplicate the environment of a face-to-face meeting. In a two-way, full-motion configuration, for example, the participants can see and hear each other.

Nonmotion videoconference: A nonmotion videoconference supports the relay and exchange of still pictures.

Personal communications networks (PCNs): As defined in the chapter, a family of mobile or portable radio communications services designed to meet the communication requirements of people on the move.

Personal videoconferencing: Personal videoconferencing is a general term that describes desktop videoconferencing systems.

Telecommuter: An individual who works at home and maintains contact with the office by computer.

Teleconference: An umbrella term for the various categories of electronic meetings and events, ranging from audioconferences to videoconferences.

Videoconference: A teleconference wherein, as implied by the name, video or visual information is exchanged.

Afterword

An afterword can serve as a footnote, a way to wrap up the topics presented in the body of a book. This afterword, though, has two functions.

The first is that of a traditional afterword. Both current and future trends are outlined. The second is to serve as a brief introduction to *virtual reality* (VR). It is one of the more interesting and powerful technologies, with an array of applications.

GENERAL TRENDS

As described throughout the book, the new technology universe is constantly expanding and is in a state of flux. New technological advances lead to new applications and, possibly, to a wide range of implications.

We are also on the verge of important developments. HDTV programming may be relayed to us by local television stations while optical media promise to alter forever the way we work and play. Similarly, new high-speed communications lines will cut across the country as personal communications services continue to span the globe.

These are only a few of the trends that may continue to shape our information society, and in the context of this discussion, others can be identified, including the following.

1. On the political front, an NPN-type network could nurture an electronic democracy while simultaneously providing access to vast information resources. CD-ROMs and other optical media, including recordable configurations, will serve as information- and electronic-publishing sources in their own right.

2. PC developments should continue to blossom, and new systems may sport GUIs and other interfaces, including a VR interface, that will make the human–computer link even more transparent. Speech recognition may also play a prominent role in this field, and the application could be extended to television sets and other everyday appliances.

3. The convergence between technologies will keep pace, and the resulting synergy should prove to be fertile ground for new products and applications. One such area is the convergence between the computer and audio-video fields (e.g., editing, and in the general communications field, interactive television systems). As described in Chapter 11, we may also witness the further integration of computers and personal communications systems.

Entire communications systems may be similarly influenced. We may witness the development of an integrated information and entertainment highway, be it by the telephone or cable industries or both. These systems, which could deliver conventional television programming, may also serve as bridges to information resources ranging from computer data to multimedia presentations.

4. Digital technology will continue to play a key role in our future communications system, as attested to by the many faces of digital video. New communications lines and other digital developments ranging from image-manipulation techniques to compression schemes will similarly affect our lives.

5. The personal-media universe will continue to expand. By using desktop publishing and video systems, for example, we become producers and not just consumers. When combined with intelligent controllers, the new media may support personal narrowcasting, as described in Chapter 8.

This facet of the communication revolution cannot be overstated. We now possess a remarkable set of personal communications tools. You can let your imagination soar as you create a 3-D graphic, and you can gain access to tailored pools of information and entertainment programming.

By extension, our communication options

Figure A-1
This drawing highlights the interrelationships between photonics technology and its various fields. As stated in the caption, "photonics is the technology of generating and harnessing light and other forms of radiant energy whose quantum unit is the photon. The range of applications of photonics extends from energy generation to communications and information processing," some applications of which have been covered in previous chapters. (Courtesy of Photonics Spectra; *Diane L. Morgenstein in* Photonics Spectra*)*

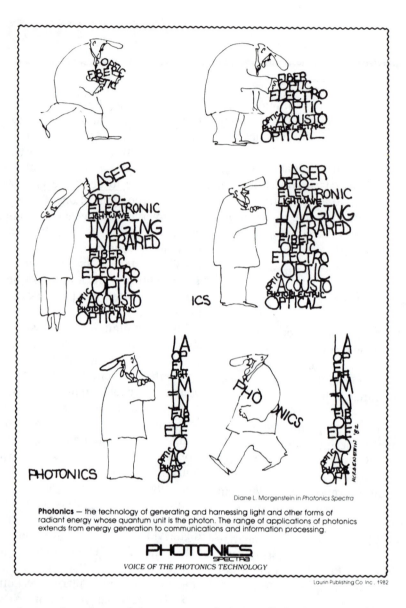

Diane L. Morgenstein in *Photonics Spectra*

Photonics — the technology of generating and harnessing light and other forms of radiant energy whose quantum unit is the photon. The range of applications of photonics extends from energy generation to communications and information processing.

PHOTONICS
SPECTRA
VOICE OF THE PHOTONICS TECHNOLOGY

Laurin Publishing Co. Inc., 1982

should continue to broaden and deepen, especially in the area of wireless communications systems. We may use personal communications devices and readily exchange data via a PC in the confines of a building or even from the field. In other applications, we may hold a videoconference from either the office or our home, while for more people the home may, in fact, become an office.

6. Ethical and legal questions will loom ever larger. Pertinent topics include colorization, digital image manipulation, and sampling.

Concerns raised over possible privacy vio-

lations will also be addressed. As described in different chapters, the right to privacy has been challenged by the new technologies. Your E-mail may be read and the government could impose new restrictions that would open up a communications system to electronic eavesdropping.

The First Amendment and its attendant rights will also continue to be challenged and defined for the electronic media. Journalists, for their part, may help to shape these fields as well as the satellite industry with respect to remote-sensing applications.

7. Information will remain an important

resource. As our information society matures, the production of and the demand for information will increase.

We will also develop new ways to manipulate and use this information. The pertinent tools range from extensions of the communications systems currently in use to the potential adoption of VR and newer configurations for business data analysis and recreational applications, among others.

But despite the growth of the information sector, it's also important to remember that heavy industries are still vital. Basically, someone has to make the steel and other products we use daily.

The information age has produced new tools that can streamline and control various industrial operations, as may be the case with computer automation or possibly a CAD program. Yet from a systems perspective, these information tools, even though they may be intrinsically linked with heavy industry, are only a part of the picture. The integration of and the correct balance between an information and industrial base could help propel a country toward the future. If the balance is lost, a country's economy could suffer.

8. Another issue is one raised in the teleconferencing chapter. Some individuals believe the new technologies may be too much of a good thing. In essence, will the electronic clutter created by the diverse assortment of communications tools become so pervasive that these tools become more of a distraction than an aid? How do you strike a balance between the need for quiet, reflective periods and your communications and information needs?

A related issue touches on potential health issues (for example, radiation from PC monitors) raised by the new technologies.

9. While the current and projected applications are exciting, they must also be examined with a dose of reality. One of the problems with predictions is that they often remain predictions even after a number of years. In one classic example, artists and authors in the 1950s and 1960s predicted or envisioned the future of the space program. Moon bases would be sprawled over the lunar surface, and space stations, much like

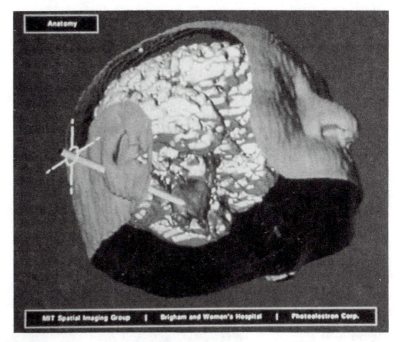

Figure A–2
New imaging techniques will have an impact on many disciplines, including the medical field. This shot, with another image, depicts the simulated radiation treatment of a brain tumor. In this particular view, the spatial relationships between the probe and brain structures, including the tumor on the brain stem, are clearly revealed. (Photo and captions courtesy of and © MIT Media Laboratory 1992; Magnetic Resonance Image, Stephen Benton's research, Spatial Imaging)

the one portrayed in *2001: A Space Odyssey,* would orbit the Earth.

As of the early 1990s, many of these predictions have not materialized for a variety of reasons. Part of the problem was money, or the lack of it, and another part may have been the technical and/or political inability to transform these visions into a reality.

The same sequence of events could affect the new communications technologies. While we may have the capability to develop an NPN-type network, the question remains, will we? While computers can serve as educational tools, will schools be equipped with enough units to make a difference?

VIRTUAL REALITY

On the Threshold of a Dream

As indicated in the opening, the Afterword wraps up with a brief overview of VR. The topic is included at this point since it is a footnote to many of the technologies and applications described throughout the book. These range from computers to teleconferencing to information processing. The implications similarly cut across social and ethical lines.

In brief, VR can be defined as follows. Virtual Reality is a display and control technology that can surround a person in an interactive computer-generated or computer-mediated virtual environment. Using head-tracked head-mounted displays, gesture trackers, and 3D sound, it creates an artificial world of visual . . . and auditory experience. With a digital model of an environment, it creates an artificial place to be explored with virtual objects to be manipulated.[1]

As just described by Michael W. McGreevy, one of the field's pioneers, you can use a VR system to enter into and to interact with a computer-generated environment. The passage to this realm, also referred to as cyberspace, can be through various VR platforms.[2] This includes the configuration we are probably the most familiar with, where you wear special "goggles" to view the computer-generated images as well as a special glove to navigate through and interact with this environment.[3] Realistic sound may also be a key element, depending upon the application, while head motions are tracked so the scene shifts as you move your head.

When you are outfitted with this array of equipment, the real world is typically blocked out as you are fed images of a room, for example. You can move through this space with a gesture of your hand. You can also grab and move objects and even turn wall switches on and off. The objects can act as

Figure A–3
A VR outfit. (Courtesy of NASA, Ames Research Center)

they would in a real-life situation, but in this case, the action is taking place in a virtual world, the world generated by the computer.

Applications

Even though VR and its technical foundations have existed for a number of years, the early 1990s saw its full emergence on the public stage in books, magazine articles, and television reports. As of this writing, a sampling of applications includes the following.

1. *Telepresence*: Telepresence is the "ability to interact in a distant environment through robotic technology."[4] In one example, NASA is experimenting with remote robotic systems, controlled by distant operators, for various space-based construction and repair operations. Rather than simply pressing buttons to carry out commands, the robotic device becomes a natural extension, of sorts, of its human operator.

2. *Teleconferencing*: With conventional teleconferencing you may only see another individual on a monitor. In a VR setup, though, your virtual image or self, a figure you see with your goggles, could shake hands with your counterpart's virtual image. You could also meet in the environment of your choice, be it New York City or a tropical beach.

3. *Architecture*: You could walk through a building before it is constructed, so you could judge the design's effectiveness and, if necessary, make modifications. This capability may be one of the sensations closest to living in a building, to try it out, before it is built. A VR system could also be used to help design buildings that better accommodate the needs of disabled individuals (for example, someone in a wheelchair).[5]

4. *Medicine*: There are numerous medical applications. A surgeon could practice a procedure with a virtual patient; a medical student could literally take a tour through the human body.

5. *Education*: Educational applications could include the medical tour, meeting historical figures, and walking through an ancient city or even on the surface of another world.

6. *Personal Freedom*: The last statement

Figure A–4
An example of a scene you can now generate and, more important, explore with your PC: Olympus Mons on Mars. (Software courtesy of Virtual Reality Laboratories; Vista Pro)

brings us to a potent VR area. In a virtual environment, you could assume a new identity or explore environments you may never be able to experience in real life.

NASA, for instance, is already using data generated from its outer-space probes to make it possible to explore other worlds in the solar system. While scenery-generator programs for PCs currently exist, where a realistic scene is created and displayed on a monitor, a VR-based system would push this application to a higher level (for example, as you turn your head, the scene realistically shifts).

7. *Other Applications*: Other application areas include military war games, enhanced scientific visualization operations, art, and virtual sex. A VR system could also serve as a sophisticated and natural human–computer interface, especially if it incorporates speech-recognition capabilities.

Computer games are similarly served. A VR system could support the ultimate video-type game. You, or your virtual self, literally become part of the action. Users received a taste of this potential through Mattel's Power Glove for the Nintendo system, and VR arcades have already been launched.[6]

8. *Implications*: As is the case with other topics discussed in the book, VR applications raise a number of ethical and social questions.

In one example, a VR configuration could be used for torture. In another example, some individuals have indicated that VR may have a powerful addictive and "intelligence-dulling" power, much like, as has been claimed, contemporary television.[7]

Other concerns have touched on the potentially insulating quality of a VR system. If we use realistic simulations, could we become desensitized to a real war in the real world where people actually die? In essence, could war turn into a VR game, much like many of the bombing runs in Operation Desert Storm, as portrayed on television, took on the appearance of a videogame?[8]

In another situation, will people be able to strike a balance between the virtual world and the real world? Other questions and concerns exist as well.

Summary

Like other technologically driven fields, the tools that constitute the VR universe should continue to mature. The equipment should become lighter, more mobile, and less physically hampering. The graphics quality should also improve, to help counter one contemporary complaint, while developments in feedback systems should keep pace, for example, providing you with a physical response similar to the one you may experience when, for example, you pick up a real ball.

Building upon contemporary systems, a room, much like the holodeck in "Star Trek: The Next Generation," could also be constructed. In this setting, you would be unencumbered by physical equipment, and as portrayed in different episodes of the television series, a fine line may be drawn between reality and what is a virtual reality.[9]

Finally, while the number of potential VR applications is enormous, the field's growth and widespread acceptance is still an unknown element at this time. But if VR systems do become a common fixture, we may all be, as stated in the Moody Blues' album title that opened this chapter section, on "the threshold of a dream."

CONCLUSION

As indicated, the new-technology universe is a dynamic one. Products are entering the market at a dizzying rate, new implications are raised, and we must learn how to manage these new tools and their fallout.

One possible solution is to become active participants and students of this venture. This holds true whether we are in school, work for a television network, or use the information provided by a hologram for independent research.

In a similar vein, it's also important to remember that the communications field extends far beyond the boundaries of the traditional radio, television, film, and cable industries. It also embraces a wide array of interesting and rapidly changing technologies that help make our world equally fascinating and dynamic.

As a reflection of our new communications era, we have viewed distant worlds and the bottom of the oceans' floors through the pictures relayed by our outer-space probes and undersea explorers. We have, in fact, witnessed sights about which previous generations have only dreamed.

Finally, we should keep in mind that the communication revolution is not merely one of technologies. The technologies simply provide us with the tools; we must decide how to use them. Think of the communication revolution as a new opportunity to explore our world. In the right hands, a technological revolution can bring about a revolution of the mind, imagination, and human spirit.

REFERENCES/NOTES

1. Michael W. McGreevy, "Virtual Reality and Planetary Exploration," paper, 29th AAS Goddard Memorial Symposium, March 1991, Washington, D.C., 1.
2. Cyberspace is a term coined by William Gibson in his book *Neuromancer*. The term also has connotations beyond our present discussion.
3. "The Future Is Now," *Autodesk*, 1992, 32.
4. Linda Jacobson, "Virtual Reality; A Status Report," *AI Expert* 6 (August 1991): 32.
5. Ben Delaney, "Where Virtual Rubber Meets the Road," *AI Expert; Virtual Reality 93: Special Report*, 1993, 18. Please note: This special edition of *AI Expert* has a number of articles that cover the spectrum of the VR field.
6. Howard Eglowstein, "Reach Out and Touch Your Data," *Byte* 15 (July 1990): 283–90;

other articles and application notes describe how to interface the Power Glove to a PC.

7. Howard Rheingold, *Virtual Reality* (New York: Summit Books, 1991), 355.
8. See Rheingold, *Virtual Reality,* 357–62, for some of the implications.
9. Tom Reveaux, "Virtual Reality Gets Real," *New Media,* January 1993, 34.

ADDITIONAL READINGS

There is a rich assortment of material covering the VR field, including articles in magazines, such as *Verbum* magazine, and the following sample titles.

Aukstakalnis, Steve, and David Blatner. *Silicon Mirage.* Berkeley, Calif.: Peachpit Press, Inc., 1992.

Benedikt, Michael, ed. *Cyberspace: First Steps.* Cambridge, Mass.: The MIT Press, 1991.

Ellis, S. R. "Nature and Origins of Virtual Environments: A Bibliographical Essay." Reprint. *Computing Systems in Engineering* 2 (1991): 321–47.

Krieg, James C. "Accuracy, Resolution and Latency: VR's Most Misunderstood Terminology for Electromagnetic Tracking." Paper. *Electronic Imaging West 92,* Anaheim, Calif., March 24-26, 1992.

Lavroff, Nicholas. *Virtual Reality Playhouse.* Corte Madera, Calif.: 1992. (The book includes 3-D glasses and demo programs.)

Rheingold, Howard. *Virtual Reality.* New York: Summit Books, 1991.

Wenzel, Elizabeth M. *Three-Dimensional Virtual Acoustic Displays.* NASA Technical Memorandum 103835, July 1991.

Wheeler, David L. "Computer-Created World of 'Virtual Reality' Opening New Vistas to Scientists." *The Chronicle of Higher Education* 37 (March 13, 1991): A6, A12, A13.

Wylie, Philip. *The End of the Dream.* New York: DAW Books, Inc., 1973. See pp. 162–65 for a precursor look at a virtual sex-type setup. The book, a science-fiction novel, primarily focuses on environmental issues and the Earth's potential future.

Index